Mentoring Mathematics Teachers

Designed to support both teachers and university-based tutors in mentoring pre-service and newly qualified mathematics teachers at both primary and secondary levels, *Mentoring Mathematics Teachers* offers straightforward practical advice, underpinned by the latest research and theory.

Developed by members of the Association of Mathematics Education Teachers, themes covered include:

- Recognition of the importance of pedagogical content knowledge
- Building upon subject knowledge
- Developing skills of self-evaluation in order to reflect and develop practice
- The on-going need to address issues of equity and diversity within the profession
- The need for beginning teachers and their mentors to work together effectively as a partnership
- The importance of collaboration, shared goals, mutual benefit and growth.

Addressing issues of mentoring for all pre-service and beginning mathematics teachers, *Mentoring Mathematics Teachers* demonstrates both the importance of mentoring in the development of new teachers of mathematics, but also the benefits to all those who involve themselves in this challenging and rewarding task.

Rosalyn Hyde is Senior Teaching Fellow at the University of Southampton, UK, Associate of the National Centre for Excellence in the Teaching of Mathematics (NCETM) and Senior Fellow of the Higher Education Academy, UK.

Julie-Ann Edwards is Lecturer in Mathematics Education at the University of Southampton, UK, and Senior Fellow of the Higher Education Academy, UK.

Mentoring Mathematics Teachers

Supporting and inspiring pre-service and newly qualified teachers

Edited by Rosalyn Hyde
and Julie-Ann Edwards

Routledge
Taylor & Francis Group
LONDON AND NEW YORK

First published 2014
by Routledge
4 Park Square, Milton Park, Abingdon, Oxon OX14 4RN
605 Third Avenue, New York, NY 10017

Routledge is an imprint of the Taylor & Francis Group, an informa business

British Library Cataloguing in Publication Data
A catalogue record for this book is available from the British Library

Library of Congress Cataloging-in-Publication Data
Mentoring mathematics teachers : how to support and inspire pre-service and newly qualified teachers / edited by Rosalyn Hyde and Julie-Ann Edwards.
pages cm
Includes bibliographical references and index.
1. Mathematics–Vocational guidance. 2. Mathematics teachers–Counseling of.
I. Hyde, Rosalyn, editor of compilation. II. Edwards, Julie-Ann, editor of compilation.
QA10.5.M46 2013
510.71'2–dc23
2013012643

ISBN: 978–0–415–81989–3 (hbk)
ISBN: 978–0–415–81990–9 (pbk)
ISBN: 978–1–315–88627–5 (ebk)

Typeset in Galliard
by Keystroke, Station Road, Codsall, Wolverhampton

Contents

Notes on contributors

Jenni Back is an experienced teacher and mathematics educator. She has taught mathematics at all levels from Early Years to A level and is passionate about developing students' understanding of mathematics and helping them to think and reason mathematically, particularly at primary level. Her research has focused on this and on ways of supporting teachers to do this. She is an experienced teacher educator and resource developer having worked for Plymouth and Middlesex Universities and the NRICH project at the University of Cambridge. She is a member of the Mathematical Association and Association of Teachers of Mathematics and is currently Chair of the Association of Mathematics Education Teachers.

Patrick Barmby is a lecturer in Primary Mathematics at Durham University and the Programme Director for the BA Primary Education degree there. In the past, he has published on a broad range of areas, including attitudes towards science and teacher recruitment and retention. However, his main areas of research are the notion of understanding in mathematics, the role of representations in understanding and reasoning in mathematics, and teacher subject knowledge in mathematics. He has recently carried out a research project looking at developing the use of visual representations in the primary mathematics classroom.

Laurinda Brown taught mathematics for 14 years at the same school, progressing to head of department. During those years she acted as mentor for the Universities of Exeter, Bath and Bristol at various times. She now works at the University of Bristol, Graduate School of Education on the one-year Postgraduate Certificate in Education course. She has written and edited widely and current research interests include developing as a mathematics teacher, developing as a mathematics teacher educator, and complex decision-making in the classroom, all informed by an enactivist perspective on learning.

Ian Campton taught in a wide variety of rural and inner-city primary and junior schools for twenty years, becoming an assistant headteacher. Much of this time involved the role of mathematics co-ordinator and manager, supporting the development of the teaching and learning of mathematics. He currently works at the University of Southampton where he contributes to programmes of initial teacher education, including mathematics education on the Primary PGCE and Secondary PGCE programmes. His main research interests include the mathematical subject knowledge of primary school teachers and how this can be developed through programmes of continuing professional development that lead to sustainable changes within their individual practices.

Alf Coles taught secondary mathematics in London and Bristol comprehensive schools for 15 years, becoming a Head of Mathematics and latterly an Assistant Headteacher. He completed a Masters in Education at the University of Bristol and then a doctorate, sponsored by the Economic and Social Research Council. In 2010, he took up a position at the University of Bristol teaching on the Postgraduate Certificate in Education course. His research interests include investigating the roles of the mathematics teacher and the facilitator of teacher learning and, in particular, aspects of these roles linked to listening and *meta*-communication.

Julie-Ann Edwards has extensive experience over more than 20 years of teaching mathematics across the 5 to 18 age range in multi-ethnic schools in New Zealand and England. She is currently working in the Southampton Education School at the University of Southampton where she teaches on the primary and secondary PGCE programmes and works with practice-based mentors on Masters and PhD programmes. Her research interests include learning mathematics in collaborative settings, the impact of friendship groupings on affective relationships in learning mathematics and the professional development of mathematics teachers, including the role of reflective practice in teacher development.

Ruth Edwards is a highly experienced teacher, having many years' experience teaching mathematics in secondary and post-16 institutions. She has taught across a wide range of mathematics programmes at all levels. She currently works at Southampton Education School at the University of Southampton where she leads the mathematics subject knowledge enhancement course as well as working with students on the secondary PGCE. Ruth works closely with the Further Maths Support Programme through which she delivers a range of student enrichment activities as well as teacher CPD. Her research interests include the transferability of mathematics to outside the classroom and applied and practical approaches to learning.

Catherine Foley was a primary school teacher in West Berkshire and Oxfordshire and subsequently became a leading mathematics teacher and local authority consultant. After several years working for the Primary National Strategy as a senior adviser in the mathematics team she moved to her current role as mathematics tutor in initial teacher training at the University of Reading, working across a range of undergraduate and postgraduate programmes. Her research interests include the ethnographic study of children's perceptions of mathematics, and she continues to work closely with schools providing CPD for teachers and support for parents.

Taro Fujita is a lecturer in mathematics education at Exeter University. He was trained as a primary and secondary school teacher in Japan, and completed his PhD at the University of Southampton. He has almost 10 years of experience working in the field of mathematics education, first at Glasgow University (BEd) and then at Plymouth University (BEd and PGCE). His current research interests include the history of mathematics education, the teaching and learning of geometry in lower secondary schools, deductive reasoning and intuitive skills in geometry and the use of technology in mathematics education.

Ray Huntley taught in several English and one Australian primary school for eighteen years and was a headteacher in Essex. He has taught on undergraduate and postgraduate courses in Teacher Education at three UK universities, delivered INSET in mathematics education to schools across the UK and lectured on mathematics enhancement courses in the UK and South Africa. He is an active member of mathematics subject associations and currently works as Lecturer in Education at Brunel University. His research is on the choices of mathematical examples by pre-service primary trainees.

Rosalyn Hyde taught mathematics in Hampshire schools for 10 years before joining the Southampton Education School at the University of Southampton. She leads the secondary mathematics PGCE course as well as contributing across other programmes in the school. She has also previously worked as Professional Development Officer for the Mathematical Association and has worked on a range of professional development programmes as a consultant and writer. Her research interests are in the development of pre-service teachers, developing pedagogic subject knowledge and the use of digital technologies in the teaching of secondary mathematics.

Manjinder Jagdev taught secondary mathematics in Nottingham, Bristol and Norwich comprehensive schools for ten years, including inner-city predominantly white schools and multi-ethnic schools. She became a school teacher mentor for University of East Anglia PGCE, head of department and UEA PGCE mathematics tutor. In addition to her current work with PGCE students in Sheffield, she has taught PGCE students at St. Mary's University College in Twickenham and at the University of Manchester. Her unpublished reports include 'Educational experiences of Black and Asian pupils in an inner-city multicultural comprehensive'. Her research interests are based on activities for classroom mathematics from many cultures.

Keith Jones works at the University of Southampton where he is deputy director of the university's Mathematics and Science Education Research Centre. Prior to this he taught mathematics for more than ten years in a number of multi-ethnic inner-city comprehensive schools, rising to become head of department. His main research interests focus on the teaching of geometrical and spatial thinking in mathematics, and on the integration of technology. As well as being a mentor when a head of department, he has considerable experience of working with mentors and trainee teachers through his roles on the PGCE programme at the University of Southampton.

Rachel Lofthouse is Head of Teacher Learning and Development (Education) and Director of the MEd in Practitioner Enquiry and PGCE tutor at Newcastle University. She is also Partnership Development Director for the Research Centre for Learning and Teaching (CfLaT). Her research, consultancy and teaching relate to developing innovative classroom practice through enhanced teacher learning. She led a two-year project funded by CfBT and the National College on Improving Teacher Coaching in Schools. Related interests include teachers' use of video for self-reflection, frameworks for lesson observation and dialogue in support of professional learning.

Avril Mallett has taught secondary mathematics for over thirty years in Norfolk after training in multicultural schools in London. She is currently second in faculty with an emphasis on Key Stage 3 at Sprowston Community High School, one of the largest comprehensive schools in Norfolk. She has rewritten schemes of work so they contain spiritual, moral, social and cultural values, ensuring that multicultural and other activities are incorporated in Key Stage 3 lessons. She has been a teacher mentor to University of East Anglia trainees for over eight years. She is interested in developing spiritual, moral, social and cultural values in junior school mathematics.

Mark Simmons began his teaching career as a physics teacher in Nottingham in 1985. Over twenty years in classrooms he has served as a head of year, head of physics, primary teacher and mathematics Advanced Skills teacher before joining the University of Nottingham School of Education in 2005 as a tutor in mathematics education. He is now link tutor for mathematics and is completing his Masters in Education. His interests are varied and include the effects of elitism, the didactics of mathematics, and teaching adults to teach.

Mary Stevenson taught secondary mathematics in the North West of England for eleven years, including four years as head of department and PGCE mentor. She joined the Mathematics Education team at Liverpool Hope University in 1997. Her teaching at Hope encompasses PGCE Mathematics, undergraduate degree mathematics, mathematics Subject Knowledge Enhancement courses, and Continuing Professional Development for serving teachers. She has been coordinator of the Mathematics Education team for ten years. Mary is fascinated by what constitutes subject knowledge for teaching, and the interplay between subject matter knowledge and pedagogical content knowledge, and this is a focus of her current research.

Geoff Tennant is an associate professor and Head of Teaching Programmes at the Institute of Educational Development, Aga Khan University based in Dar es Salaam, Tanzania, working with Masters programmes for teachers and educational leaders throughout the East Africa region. He previously spent twelve years in secondary mathematics initial teacher training at the Universities of Leicester and Reading, with experience also of school teaching in London and Berkshire. His professional and academic interests reflect a concern that all children should have maximal access to the curriculum, particularly within mathematics, with research interests in the teaching and learning of elementary algebra.

Danielle Townsend is currently a head of mathematics in a Hampshire secondary school. She completed her PGCE at the University of Southampton and was appointed as a mathematics teacher in a partnership secondary mathematics department where she became the mathematics mentor for pre-service teachers. Since then she has undertaken a part-time MSc in Mathematics Education at the University of Southampton in which she explored her interests in mentoring, particularly in mentoring potentially outstanding pre-service teachers and in the ongoing impact of collaborative paired placements.

Paul Wilson has been a mathematics teacher since 1980 and was the head of a very large mathematics department from 1991 to 2004. He has been involved in teacher education throughout his career, working as a school-based mentor for several universities and teaching on the BEd mathematics programme at Exeter University. He joined University College Plymouth St Mark and St John in 2004 and since September 2010 has been the programme leader for BEd mathematics. He has been actively involved in the development of collaborative placements for his institution since 2007. As well as developing collaborative paired placements, his recent research has focused on students' dispositions and affective responses to learning mathematics.

David Wright is programme leader for secondary mathematics PGCE at Newcastle University. He taught in secondary schools and further education for fifteen years before joining the British Educational Communications and Technology Agency (Becta) as subject officer for mathematics. He was an editor of *Micromath*, a journal of the Association of Teachers of Mathematics (ATM). He is the author of articles and book chapters on the integration of ICT in mathematics classrooms and the professional learning of teachers.

Introduction

The learning journey and lifelong learning

Rosalyn Hyde and Julie-Ann Edwards

The beginning

There's a well-thumbed slim volume in our university library called *Mentoring in Mathematics Teaching* (Jaworski & Watson 1994). It has been out on loan many times to staff and students and sometimes several borrowers want the book simultaneously. When a Master's degree student of ours started a dissertation on mentoring she found that buying her own copy of this book would be extortionately expensive because it has been out of print a long time. Her experience, along with conversations at the Association of Mathematics Education Teachers, led us to suggest that we would like to edit a new volume, inspired by this first work, on behalf of the Association. Much has changed in the almost twenty years since Barbara Jaworski and Anne Watson produced their book, so we felt that the time was right to propose the writing of a new book on the same theme. This book is the end result.

Mathematics is often thought of as a difficult subject to learn. Our contention is that it is more the case that the subject matter is complex and the job of teaching mathematics is highly skilled, although there is much research in the area of teaching and learning mathematics available to help teachers (see, for example, Watson, Jones & Pratt 2013). We argue that there are skills and abilities specific to teaching mathematics that pre-service teachers need to develop. Learning to teach mathematics is therefore a challenging undertaking, as is mentoring a pre-service or beginning teacher of mathematics. The aim of this book is to support those who undertake this mentoring role.

All the chapters in this book are jointly written by pairs of authors; in most cases, one of the authors is a specialist in primary mathematics education and the other in secondary mathematics education. Some of the authors are school-based mentors, but all have extensive experience of working with pre-service and beginning teachers and a strong desire to see learners in schools experience high-quality mathematics teaching.

The importance of mentoring

The importance and value of practice-based placement experience in the preparation of new teachers has never been in doubt. However, teaching school mathematics is complicated – there is much that could go wrong and many variables to consider. Learning by trial and error is a risky undertaking in such a context so the guidance, support and wise advice of a mentor is vital in providing pre-service teachers with appropriate learning experiences so that they develop effective skills as teachers of mathematics. Reading the chapters in this book we are struck by the skill required to mentor well, which is especially notable when teachers are mentoring alongside their main role as a classroom teacher of mathematics.

Themes

There are four recurring themes in the resulting book, appearing in one guise or another in almost every chapter. The first of these is the recognition of the importance of pedagogical content knowledge for mathematics teachers. There is clear understanding that this is not the same as personal mathematical knowledge and that pre-service teachers have much to learn about how mathematics is learnt and how it can be taught. The Initial Teacher Education Matters report (National Centre for Excellence in the Teaching of Mathematics 2011: ITE2) also identifies that effective initial teacher education utilises "effective mathematical pedagogical knowledge, judgement and practice developed through informed understandings of relationships between procedural and conceptual knowledge."

A second theme running through this book is that of reflective practice. The importance of pre-service teachers developing skills of self-awareness and self-evaluation in order to reflect and develop their own practice with the intervention of a skilled mentor cannot be over-emphasised. Again, this is emphasised in the Initial Teacher Education Matters report where the exemplification suggests that pre-service teachers can engage in reflective practice through:

- taking time to pause, to critically evaluate, to reflect and consider, to consolidate and accommodate what they have been learning;
- developing reflection as a strategy for supporting their own professional development;
- shared reflection with peers and others;
- embedded approaches to reflection and reflective practice throughout their professional life;
- reflection playing a part in the process of reconceptualising and reconstructing their mathematical knowledge and developing teacher identity;
- . . . being able to take notice of what they've observed and implementing it so that it has an impact on their practice.

(National Centre for Excellence in the Teaching of Mathematics 2011: ITE3)

Underpinning much of the writing in this book is a third theme: that beginning teachers and their mentors work together effectively when they function as part of what Lave & Wenger (1991) called a 'community of practice'. In such communities, it is acknowledged that the pre-service teacher and mentor are part of a group that share a common interest (probably that of teaching mathematics well) and work together to share practice through joint activities and discussions that facilitate learning from one another.

A final theme of the book is that of collaboration. In every chapter there is an emphasis on purposeful and productive working together towards shared goals. This goes further than mentor supporting beginning teacher because it embodies a sense of co-learning and mutual benefit and growth.

The four themes together demonstrate both the importance of mentoring in the development of new teachers of mathematics, but also the benefits to all those who involve themselves in this challenging and rewarding task.

The learning journey

In the Initial Education Matters consultation, many of our colleagues working with pre-service teachers shared the view that becoming a mathematics teacher should be seen as a lifelong learning journey with a continued need to learn and develop as a teacher through professional development and by responding to new developments, technologies and research. The exemplification in the report goes on to suggest that pre-service teachers can undertake this by:

- recognising that as teachers ITE trainees (pre-service teachers) continue to develop throughout their career.
- becoming involved and contributing to collaborative learning communities.
- developing the skills to become the subject leaders, leading practitioners and innovators of the future.
- recognising that becoming a teacher is a lifelong commitment to their own learning.
- becoming resilient in the face of change, able to acquire new skills, evaluate and innovate in response to new technologies.
- keeping informed of new developments and research.

(National Centre for Excellence in the Teaching of Mathematics 2011: ITE11)

Inherent in this book is an implicit understanding that 'becoming a teacher' is not the end of anything; initial qualification might be more realistically seen as one of the 'staging posts' in a career in teaching, sometimes with an exciting ride in between the posts, hence the notion of a 'learning journey'.

The analogy of a 'learning journey' is commonly used in educational contexts and, in particular, in initial teacher education by those who use reflective models of learning. Such reflective models provide a means to link the theoretical and practical aspects of learning and to describe the process whereby those who have been students of mathematics become teachers of the subject. Such an analogy also provides a means of using further analogies to explicate learning, for example, "Recognising landmarks along a personal learning journey serves to confirm one's direction" (van Halen-Faber 1997: 59).

Conway and Clark (2003) further describe this learning journey as being both 'inward' and 'outward'. They describe the 'outward journey' as taking pre-service teachers through the three stages of what is known as Fuller's model of novice teacher development with: "Stage One: concerns about self; Stage Two: concerns about tasks/situations; Stage Three: concerns about impact on students" (Conway & Clark 2003: 466) and on the 'inward journey' pre-service teachers "engaged in a progression toward greater self-awareness/self-knowledge and subsequently made efforts at greater self-organisation and self development" (*ibid.* p. 470).

It is our hope and intention that this book will encourage and support all those who work with pre-service teachers and early-career teachers to continue both on their own learning journeys as well as influencing the journeys of others.

References

Conway, P. & Clark, C. (2003) 'The journey inward and outward: a re-examination of Fuller's concerns-based model of teacher development', *Teaching and Teacher Education*, 19, 465–82.

Jaworski, B. & Watson, A. (1994) *Mentoring in Mathematics Teaching*, London: Falmer Press.

Lave, J. & Wenger, E. (1991) *Situated Learning: Legitimate Peripheral Participation*, Cambridge: Cambridge University Press.

National Centre for Excellence in the Teaching of Mathematics (2011) *Initial Teacher Education Matters*. Online. Available at HTTP: <https://www.ncetm.org.uk/resources/33510> (accessed 21 December 2012).

Van Halen-Faber, C. (1997) 'Encouraging critical reflection in preservice teacher education: a narrative of a personal learning journey', *New Directions for Adult and Continuing Education*, 74, 51–60.

Watson, A., Jones, K. & Pratt, D. (2013) *Key Ideas in Teaching Mathematics*, Oxford: Oxford University Press.

Part I

Developing subject knowledge

Chapter 1

Partnering theory and practice in mathematics teaching

Ian Campton and Mary Stevenson

The relationship between the quality of mathematics teaching and the subject-related knowledge of the teacher is widely acknowledged as an important one. Also, the process of teaching stimulates a growth and reconstruction of knowledge, so this relationship is two-way. In this chapter we explore ways that the content knowledge of pre-service teachers can be developed in practice, and the role practice-based mentors have in facilitating these processes. The importance for practice-based mentors and pre-service mathematics teachers of nurturing a strong focus upon the subject will be explored. In placement contexts characterised by multiple competing agendas, we believe this is essential in the ongoing improvement of mathematics teaching at all levels.

Introduction

Many analogies have been used to depict the ways in which students develop an understanding of mathematics. One of them, the 'learning landscape' as constructed by Twomey Fosnot and Dolk (2001), describes how students negotiate the terrain of an area of mathematical learning in order to reach the mathematical understanding that lies on the horizon. The learning journey involves visiting the landmarks which are the key ideas and concepts behind the topic necessary for a comprehensive understanding of the mathematics that lies on the horizon. Roaming aimlessly around this landscape is a futile learning experience for students and something that should be avoided. A guide, in the form of a mathematics teacher, is therefore required to assist in their passage from the starting place to the horizon. If this is to be a smooth passage, the teacher, as with all good tour guides, needs to maintain an intimate knowledge of the landscape and the many ways in which it can be successfully negotiated. Not only is this learning terrain complex, dotted with misconceptions and barriers to learning, but the students exploring it all have different starting points and will probably take different routes. Consideration must therefore be given to the knowledge that equips the pre-service teacher for the task of teaching mathematics. Research findings in this area offer the mathematics teaching community at all levels some powerful and transformative ideas.

Attention to these ideas helps teachers to become more expert in the field and able to make informed judgements about government initiatives and other competing agendas. Ultimately, attention to the findings of research will lead to ongoing improvement in mathematics teaching for all. Much has been written about the development of mathematics subject knowledge for teaching, and yet there remains a clear mismatch between the public, policy maker and layperson's views of what subject knowledge means, and the subtler, more complex characteristics gained from research. Is the possession of a good grade in secondary school mathematics sufficient to equip someone to teach primary school students? Is the possession of a degree in mathematics sufficient to enable someone to teach mathematics in a secondary school? Most of those who work in education would agree that whilst formal academic qualifications are important, they are not sufficient alone for successful teaching. Teachers need a wider, more nuanced, specialised knowledge of the subject, which relates to the curriculum and the needs of their students. It is not sufficient to 'know' some mathematics oneself: one needs, amongst other things, to be able to 'unpack' the ideas in order to make them accessible to learners, to be able to make careful choices of examples and tasks, and to be able to respond flexibly to the responses of students to those tasks. So what exactly does this specialised knowledge consist of, how is it constructed and, perhaps more importantly for mentoring pre-service teachers, how is it developed?

What do we mean by subject knowledge for teaching?

The field of mathematics subject knowledge for teaching has been widely researched, and various models have been proposed in defining its contents and structure. Inevitably there are similarities between these models, making the sorting into discrete categories rather imprecise. With overlaps and blurred edges, an awareness by both mentor and pre-service teacher of the subject knowledge for teaching should not necessarily concern the organisation of these different categories, but rather their content, which is perhaps more important and beneficial to improving practice.

Similarities in the research field can be attributed to the fact that much recent research into knowledge for teaching is founded upon the seminal work of Lee Shulman (1986). Shulman questioned the distinction between content and pedagogy, and asked how these are related in the process of teaching. He discussed the connections between knowing and teaching, probing how learning for teaching might occur, and noted that "mere content knowledge is likely to be as useless pedagogically as content-free skill" (p. 8). He suggested that it is important to think about how these aspects of a teacher's repertoire are blended together.

In shifting the focus away from the generic aspects of teaching, Shulman proposed a three-part model of content knowledge consisting of *subject matter content knowledge* (SMK), *pedagogical content knowledge* (PCK) and *curricular knowledge* (CK). The major contribution to this model was his concept of the existence of pedagogical subject knowledge, i.e. subject matter knowledge *for teaching*. Shulman's ideas are not subject specific; they apply across the curriculum. In the context of mathematics, recognition of pedagogical subject knowledge means that there is knowledge that teachers need to have, and there are ways in which they need to hold that knowledge, which are specific to the requirements of teaching the subject and which other people who use mathematics do not necessarily have.

Subject matter knowledge

In essence, this category is concerned with detailed knowledge of the subject with such depth that it has two fundamental constituent parts as identified by Schwab (1978): *substantive knowledge* and *syntactic knowledge*. For a mathematics teacher at any level, substantive knowledge is the knowing of facts and concepts, whilst syntactic knowledge is that which is required for the process of doing mathematics. Shulman believed that teachers should not only know that something is the way it is but also why it is this way, and that they must understand why one topic is fundamental to a subject and others are secondary, knowledge he considers to be important in making pedagogical judgements. The importance of subject matter knowledge in making pedagogical judgements is something that is rarely contested, and it is a readily accepted view that teachers must have a solid knowledge of the subject they teach. But what is not so widely acknowledged is the extent of this knowledge and how deep it needs to be.

The irrelevance and relevance of formal mathematical qualifications

In returning to ideas we introduced at the start of this chapter regarding the relationship between subject knowledge for mathematics teaching and formal qualifications in mathematics, the answers to the questions posed in the introduction of this chapter need exploring. Research shows that there is no relationship between teachers' level of formal qualifications in mathematics and their effectiveness in mathematics teaching (Eisenberg 1977; Begle 1979; Askew, Brown, Rhodes, Johnson & Wiliam 1997). That is not to suggest that formal academic qualifications are not important or necessary. What this does show is that something else is going on: effective teachers of mathematics have other attributes, other knowledge areas which are often hard to pin down and measure, whilst formal qualifications are easy to quantify and record.

For the pre-service teacher and his or her mentor, this should be reassuring. For what it suggests is that developing one's subject knowledge for teaching is not really about studying high-level mathematics courses, although some pre-service and novice teachers will have done this, and enjoyed and succeeded in it; rather, it is about paying attention to the detail of the mathematics with which one is working, and endeavouring to gain a deep and connected understanding of this mathematics with regard to the needs of the learners. In particular in the context of mentoring, it is about developing openness to new ideas and being prepared to re-learn mathematics in different ways. As times change, the ways in which pre-service teachers learnt mathematics may be different to those used in the present-day classroom; some examples of this can be seen below. This can be challenging and uncomfortable for pre-service teachers, and mentors need to recognise this.

Examples of changing teaching methods

Primary

Division algorithms – the 'bus stop' method vs. the 'chunking' strategy, which is a new approach to many pre-service teachers. The 'bus stop' method is a written algorithm where the divisor is 'divided into' a uniquely partitioned dividend and remainders are 'carried over' to the next column. The 'chunking' strategy exploits the quotative aspect of division and uses a sophisticated method of repeated subtraction where 'chunks' of the divisor are removed from the dividend rather than repeatedly removing. So, for 5736 ÷ 6 using the 'bus stop method', 6 is 'divided' into 5 which does not go, so the 5 is 'carried over' into the next column to make 57. 6 into 57 goes 9 times with a remainder of 3, again 'carried over into the next column', and so the procedure continues. The 'chunking' strategy involves calculating how many 6s make up 5736 through the process

'bus stop' method

	0	9	5	6
6	5	$^{5}7$	$^{3}3$	$^{3}6$

'chunking'

		9	5	6		
6	5	7	3	6	–	
	5	4	0	0		900 × 6
		3	3	6	–	
		3	0	0		50 × 6
			3	6	–	
			3	6		6 × 6
				0		

Figure 1.1 Methods of division

of repeated subtraction. As 5736 is a large dividend, constantly taking 6 away is time consuming and so 'chunks' comprising of groups of 6 are subtracted.

Secondary

Solving linear equations – the traditional 'change side, change sign' method vs. the 'balancing' approach. For example:

> The first stage of solving the equation $3x + 4 = 19$ by the traditional procedural approach moves the +4 to the other side of the equation where it becomes –4, as shown in Figure 1.2.
>
> The 'balancing' approach would be to subtract 4 from each side in order to remove the +4 term and isolate the 3x, whilst still keeping the equation balanced, as shown in Figure 1.3.

$$\begin{aligned} 3x + 4 &= 19 \\ 3x &= 19 - 4 \\ 3x &= 15 \end{aligned}$$

Figure 1.2
Solving a linear equation by a 'traditional' approach

$$\begin{aligned} 3x + 4 &= 19 \\ -4 &\;\; -4 \\ 3x &= 15 \end{aligned}$$

Figure 1.3
Solving a linear equation by a 'balanced' approach

In a study to identify what it is that primary school teachers know, understand and do which enables them to teach mathematics effectively, Askew *et al.* (1997) found that being a highly effective teacher of mathematics was not associated with having an Advanced level or degree in the subject. As the development of basic number concepts is at the heart of the primary school curriculum, these findings are perhaps not surprising. After all, how often are primary school teachers required to draw upon either the substantive or syntactic knowledge of such higher-level mathematics and such things as the formulation of differential equations? What is important, therefore, is not so much a comprehensive knowledge and understanding of academic mathematics, but more an understanding of the mathematics within the curriculum that is taught and how critical this is to planning and implementing learning episodes. The quality of the teacher's mathematical knowledge and how this is used within teaching is of greater importance than its standing in the qualifications field. The work of Liping Ma (1999) also supports this view and highlights the importance of the quality of mathematics teachers' knowledge.

The work of Ma: profound understanding of fundamental mathematics

Liping Ma undertook a comparative study of elementary mathematics teachers in the USA and in China. She hypothesised that elementary school teachers in the two countries possessed "differently structured bodies of mathematical knowledge" (Ma 1999: xx) and that Chinese teachers might have a better understanding of mathematics than US teachers. Drawing on work by Shulman (1986) on pedagogical content knowledge, Ball (1989, 1996) on teachers' subject matter knowledge, and others, she suggested that the difference in learner performance was linked to the nature of subject matter knowledge held by teachers in the two countries, and so she investigated this by giving the teachers a set of scenarios to discuss, for example:

> Imagine that you are teaching division with fractions. To make this meaningful for kids, . . . many teachers try to relate mathematics to other things. Sometimes they try to come up with real-world situations or story-problems to show the application of some particular piece of content. What would you say would be a good story or model for $1\frac{3}{4} \div \frac{1}{2}$?
>
> (Ma 1999: 55)

The two groups of teachers responded very differently to questions and scenarios presented by Ma. The US teachers displayed a superficial level of mathematical knowledge that, when probed, was somewhat insecure. These teachers had successfully progressed through their formal education and their preparation for teaching, without really being required to understand deeply the mathematics concepts that they were handling regularly in the classroom. In contrast, the Chinese teachers showed a more secure and deep knowledge of mathematical knowledge, even though they typically had spent fewer years in formal education and training than their US counterparts. Ma describes this in-depth understanding of the mathematics at the level being taught as a '*profound understanding of fundamental mathematics*' (PUFM), and argues that this is something the successful Chinese teachers possessed and most of the US teachers did not. In the teaching and learning of subtraction, for example, the Chinese teachers possessed knowledge that not only included how to carry out a subtraction algorithm, but also why it made mathematical sense to do so.

Something that may be considered to be part of a primary school teacher's PUFM is the full and deep understanding of the number system that Askew *et al.* (1997) suggest teachers require in order to teach mathematics effectively. This is something that may not necessarily be acquired through conventional study of mathematics, and is perhaps the type of knowledge referred to by Ball, Thames and Phelps (2008) as '*specialised content knowledge*' (SCK), knowledge that enables teachers to complete mathematical work that others do not need to do.

The work of Ball in refining Shulman's subject matter content knowledge

Deborah Ball and others have been involved in research on mathematical knowledge for teaching and this has resulted, over time, in a development of the Shulman model. Ball *et al.* (2008) refined Shulman's model specifically in the context of knowledge for mathematics teaching, introducing some new categories and rethinking old categories. The Ball model retains Shulman's distinction between *subject matter content knowledge* and *pedagogical content knowledge*, but further subdivides these categories, as shown in Figure 1.4.

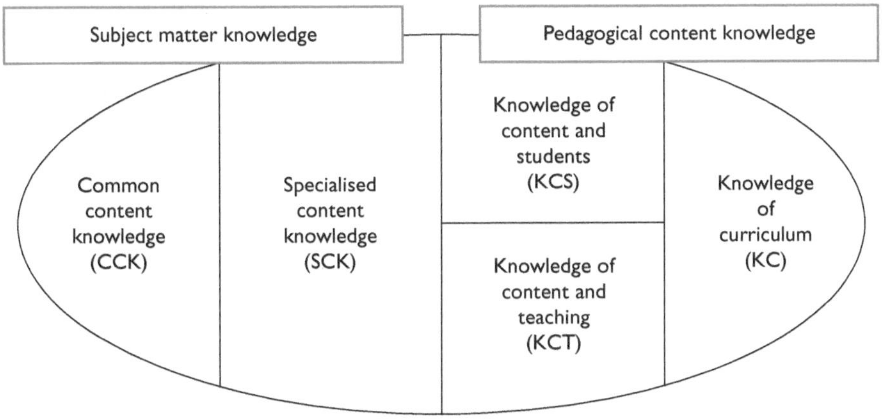

Figure 1.4 Comparison of Ball's and Shulman's categories. Adapted from Ball *et al.* (2008)

Subject matter knowledge is divided into *common content knowledge* (CCK) and *specialised content knowledge* (SCK). CCK is general mathematical knowledge that most educated adults would have. SCK, an important addition to the model, is the mathematical knowledge needed for teaching which is detailed in a way that goes beyond what is needed in everyday life and, moreover, is not necessarily known to other mathematicians. The authors argue that "the demands of the work of teaching mathematics create the need for a body of mathematical knowledge that is specialized to teaching" (Ball *et al.* 2008: 11).

An example is the knowledge and ability to analyse a piece of student's work such as a written calculation and understand the student's mathematical understandings and/or misconceptions that this reveals. Further examples can be found in Table 1.1 (see p. 14).

Ball *et al.* (2008) suggest that teachers have to uniquely unpack mathematics and hold this in order to make the content visible to students and, therefore, more learnable to them. This unpacking appears to be similar to the idea of teachers knowing the 'how' and 'why' of mathematics as suggested by Shulman (1986) and illustrated by Ma (1999).

Table 1.1 Mathematical Tasks of Teaching (adapted from Ball *et al.* 2008)

- explaining why a procedure worked and what this means;
- appraising children's computational methods and deciding whether or not any new method can be generalised;
- presenting mathematical ideas;
- responding to pupils' 'why' questions and evaluating their claims;
- finding examples and representations to make specific mathematical points or for particular purposes, recognising what is involved in using these and linking them to the underlying ideas and to other representations;
- differentiating pupils' activities;
- evaluating textbooks and adapting their mathematical content;
- posing constructive questions for pupils.

In tandem with the development of *specialised content knowledge*, Ball and her colleagues identify '*common content knowledge*' (CCK), common because it is the mathematical skills and knowledge that are used in a wide variety of settings other than teaching. Simply computing the answer to an addition calculation, for example, requires CCK.

It appears that perhaps the models of Schwab (1978) and of Ball *et al.* (2008) are similar and, in some respects, overlap. Ball's common content knowledge is similar to Schwab's substantive knowledge, as both involve the knowing of mathematical facts and knowledge for carrying out procedures. It is also difficult to see how a teacher employing Ball's specialised content knowledge does not simultaneously evoke the use of Schwab's syntactic knowledge. A major component of specialised content knowledge is appraising learners' methods for solving computational problems and determining whether such methods can be generalised to other problems. In order to do this, the teacher requires syntactic knowledge.

Pedagogical content knowledge

Pedagogical content knowledge (PCK) is the subject matter knowledge for teaching (Shulman, 1986) and is the bridge between the teacher's knowledge and enabling students to know it. Some examples are given below. Shulman's view is that PCK consists of ways in which the subject can be represented in order to be comprehensible to others and an understanding of what makes learning easy or difficult for specific topics. Within his definition, Shulman suggests three included aspects of PCK. The first is knowledge about the use of explanations, diagrams and metaphors and how these may be adjusted to suit students' learning and is referred to as KCT or 'knowledge of content and teaching'. The second, 'knowledge of content and students' (KCS), is knowledge of students' conceptions, in particular the preconceptions and misconceptions that students of different ages and from different backgrounds bring with them to their learning.

Examples of pedagogical content knowledge

Primary

Choice of examples/tasks – When teaching written calculations, 23×19 can be solved mentally without too much difficulty and therefore does not necessarily require a written calculation.

Common misconceptions – Students perceive 0.246 as being larger than 0.7 due to the number of digits it contains.

Secondary

Choice of examples/tasks – When teaching students about indices, the teacher should be aware that 2^2 is not a sufficiently general starting example since $2^2 = 2 \times 2 = 2 + 2$, and this may reinforce misconceptions.

Common misconceptions – The teacher should be aware that many learners are reluctant to offer algebraic expressions, such as $2 + 3b$, as responses to questions, thinking that the presence of an operation in the expression indicates that the task is unfinished.

Rowland and Turner: The Knowledge Quartet

Rowland, Thwaites, Huckstep and Turner (2009), in work observing pre-service teachers, have developed a staged model of teacher content knowledge known as the 'Knowledge Quartet'. It comprises four dimensions: foundation, transformation, connection and contingency. This work offers, amongst other things, a new perspective on what constitutes pedagogical content knowledge. Three of the four categories, or dimensions, of the Knowledge Quartet focus on the ways in which knowledge is applied to the teaching process as demonstrated in both the act of teaching and planning to teach. This is a different type of model from those developed by Shulman and Ball, described above, as it concerns teacher acquisition and application of knowledge rather than suggesting categories of knowledge.

The concept of the Knowledge Quartet has been developed as a framework for observing and/or reflecting upon mathematics teaching by those involved with the process; in our case, the individual beginning teachers and those mentoring them. As it has the advantage of simplicity as well as being comprehensive in covering key aspects of teacher knowledge, some schools and universities use the framework for mathematics lesson observations, and you may wish to consider this if you are not already doing so.

Foundation

The first dimension of the Knowledge Quartet, foundation, refers to the knowledge, conceptions and beliefs that beginning teachers bring with them from their own mathematical background and can be compared with Shulman's subject matter knowledge (SMK). However, as this topic has already been discussed, we would like to focus attention on the other three.

Transformation

This is the dimension in which Shulman's (1986) pedagogical content knowledge (PCK) may be observed. Transformation concerns teachers' ability to transform their own mathematical knowledge into ways that make it accessible to learners. It expands the aspect of PCK that involves knowledge about the use of explanations, diagrams and metaphors to include knowledge that enables the teacher to make the correct selection and use of equipment, mathematical representations and examples; knowledge that allows the teacher to give clear explanations of ideas and concepts, possibly with analogies; knowledge to enable accurate demonstrations of how to carry out procedures; and knowledge of assessment in order to develop students' knowledge and understanding.

Connection

Shulman's third aspect of his content knowledge model, curricular knowledge, can be observed within the connection dimension of the Quartet, which is concerned with the ordering of topics and tasks, the level of challenge, and the links made with other areas of the curriculum. This is about making decisions, informed by pedagogical content knowledge, that not only concern the links which can be made to previous lessons and within a lesson, but also concern the links between mathematical concepts and the questions that are asked in order to elicit students' understanding of these connections.

The knowledge that informs such decisions is considerable and often develops through experience. Being able, for example, to draw upon knowledge about sequencing and connectivity so that a series of lessons are coherent in their progression and are related in context and students' knowledge, is something that beginning teachers may not easily do as they do not necessarily hold such knowledge in abundance. For a more experienced teacher, creating this cohesion is often second nature and instinctive in practice.

This dimension is perhaps the most significant in its bearing on the mathematical landscape of learning. Askew *et al.* (1997) found that the ability to establish connections within students' understanding was a contributory feature to what they considered to be effective mathematics teaching. The connectionist orientation approach to teaching that emerged from their research concerns the beliefs and practices of valuing students' methods and teaching effective strategies based on their understanding. This is further exemplified in Chapter 3. According to Ma (1999) elementary mathematics is a field of depth, breadth and thoroughness, and teachers who have the same deep and thorough understanding reveal the connections between mathematical ideas rather than invent them. Connectedness is therefore one of the four qualities that Ma associates with PUFM. There is much research to support these views, highlighting the important role of the teacher in exposing these links and connections to students (see, for example, Chick & Harris 2008) and so an important part of mentoring involves mentors drawing pre-service teachers' attention to links and encouraging them to make connections within their planning and teaching of mathematics.

Another of the qualities associated with PUFM, worthy of a mention here, is something that Ma calls 'longitudinal coherence'. Its resemblance to Shulman's curricular knowledge is fairly strong as it is defined as knowledge of the whole mathematics curriculum rather than what needs to be taught to a specific age group. Developing and maintaining such knowledge is easier for beginning secondary school teachers, who regularly encounter students of different ages, than it is for beginning primary school teachers who tend to work with the same students. Either way, it is important that mentors help beginning teachers to develop and maintain this longitudinal coherence as it assists in the decisions that are made within the connection dimension.

Contingency

Contingency reflects the necessity for the teacher to respond to the needs and ideas of students and to work flexibly in order to respond to them appropriately. In essence, it is the ability to 'think on one's feet' and being prepared to deviate from the lesson plan when appropriate. It therefore requires a deep level of knowledge on the part of the teacher to be able to do this effectively. Indeed, flexibility of response to learners is often cited as the mark of an experienced teacher, and pre-service teachers are much more likely to need to adhere to a pre-prepared plan. The contingency dimension therefore takes Shulman's model beyond what we might think of as 'knowledge', and into the domain of professional skills. This idea is developed further in Chapter 8.

Responding appropriately to students' questions, answers and comments, correct or otherwise, is another skill that is likely to develop with experience and will therefore not be something that beginning teachers immediately excel in. It involves the teacher appraising the responses immediately as they happen and deciding whether they are significant in developing students' understanding, and therefore worth pursuing. Muir (2008) describes this activity as the teacher identifying 'teachable moments', the simultaneous acts that teachers make in response to students' answers, suggestions or comments and how these are used to address misconceptions or to enhance conceptual understanding. Recognising and using these moments is a crucial part in the process of allowing learners to form connections within their understanding.

Conceptual and procedural understanding

Another aspect of content knowledge that we believe to be important in the teaching and learning of mathematics is that of the relationship between conceptual and procedural understanding, referred to by Skemp (1976) as relational and instrumental understanding respectively. Conceptual, or relational, understanding is an understanding of the mathematical concepts involved. It is knowing what to do and why, and allows connections to be made between other concepts of mathematics. According to Skemp, relational understanding requires integration with other topics, demonstrating the importance of Shulman's curricular knowledge. Procedural, or instrumental understanding, is the adherence to systematic processes, rules and routines without awareness or reason. Realising that you are not actually 'borrowing one' from the 10s column as part of a written subtraction algorithm, but rather re-partitioning the number in a way that allows the calculation to be performed, is conceptual understanding. When calculating a fraction of a quantity, following the routine of 'divide by the denominator and multiply the answer by the numerator' without an appreciation of why it is procedural understanding.

There are advantages and disadvantages in both types of learning (Skemp 1976). Although conceptual understanding enables students to gain a comprehensive understanding of their learning, it can often take time to be achieved. When faced with the pressures of preparing students for tests and examinations, procedural understanding can often be favoured as a quicker teaching and learning approach. Yet once a conceptual understanding is acquired, the learning of new mathematics becomes easier, less complicated and more straightforward, saving more time and reducing the pressures. The relationship between the two types of understanding is therefore not as simple as we may think and is possibly of far greater importance in the development of students' understanding than teaching, which favours one type more than the other.

Hiebert and Lefevre (1986) maintain that learners are not fully competent in mathematics if they are deficient in either conceptual or procedural knowledge or if the two exist as separate entities. This view is supported by Rittle-Johnson, Siegler and Alibali (2001) in their research on students' understanding of decimal fractions. They conclude that conceptual and procedural knowledge appear to develop in a continuous cycle, the gains in one supporting the increases in the other, which in turn supports the increases in the first, a process that is supported by the work of others in the field (Byrnes & Wasik 1991; Rittle-Johnson & Alibali 1999; Canobi, Reeve & Pattison 2003).

Consider a student learning the multiplication tables. Initially, facts may be learnt procedurally. As a conceptual understanding develops, the student is able to make associations between the 3 and 6 times table, for example, which in turn supports the procedures through which the 6 times table is learnt.

Learning should therefore not be discussed separately in order to favour either conceptual or procedural understanding over the other (Wu 1999) and mentors need to guard against this as it is all too easy for an inexperienced beginning teacher to do just that.

How does content knowledge develop?

In the milieu of the school environment, it is absolutely vital that mentors and beginning teachers take time to talk about mathematics. This can happen both during the process of activity preparation and in discussions/feedback after teaching has taken place. It is important that discussions with beginning teachers do not focus only upon generic features (e.g. classroom organisation, behaviour management, lesson structure) but also upon the mathematics itself. For example, discussions should include how the mathematics being taught has informed choice of examples and tasks, how learners' understandings developed, and how the teacher's own understanding has changed through the process of preparing to teach this topic to new learners. A framework for supporting this is discussed below.

Research (Ma 1999) shows that in schools where teachers work together to plan and evaluate children's mathematical learning, children's mathematical attainment is good. The critical factor is making time to talk about mathematics, to always be learning while one is teaching. Ma made two important recommendations: "Address teacher knowledge and student learning at the same time" (p. 146) and "Enhance the interaction between teachers' study of school mathematics and how to teach it" (p. 147). These suggestions are based upon a belief that teacher knowledge and student learning are inextricably linked.

Watson: Mathematics knowledge as an active process – a way of being and acting

In discussing ways of finding out about teachers' knowledge, Rowland *et al.* (2009) claim that to gain a full picture it is not sufficient merely to ask the teacher to complete questionnaire items, but researchers must also observe the teacher in action. They suggest that different features of a teacher's knowledge come together "in the teaching moment" (p. 24) and are not fully accessible outside this context. In other words, teaching must actually be observed. This idea has resonance with the work of Anne Watson (2008), who sees mathematical knowledge as something which is active. She adopts a view of learning based on the idea that teachers learn through participation in practice. She discusses mathematical knowledge as "a way of being and acting" (p. 1), and she suggests that development and deepening of knowledge takes place through "doing mathematics and being mathematical" (p. 1). This is clearly supported by the ideas of Ma (see previously) who saw teachers' own use of mathematics and their successful teaching of mathematics to learners as inextricably linked.

Askew: A mathematical sensibility

Many researchers have written about mathematics subject knowledge for teaching, and the characteristics of effective mathematics teachers. We have discussed earlier the influential work of Askew *et al.* (1997) in highlighting the importance of making connections in mathematics teaching, and the links to be made with the connections dimension of Rowland *et al.*'s Knowledge Quartet. Writing more recently, Askew (2008) suggests that, rather than acquiring a body of knowledge, teachers need to develop a "mathematical sensibility" (p. 22) to enable them to handle the demands of the curriculum and to embrace change. In other words, in order to teach effectively now and into the future, teachers need to know how to learn mathematics and need to be open and confident to do this. This view of doing and using mathematics in teaching as a dynamic process is supported by Watson (see previously).

Situated learning

There is strong evidence that teacher knowledge and learning is situated (see, for example, Hodgen 2011), meaning that knowledge and learning arise from working within the particular school situation. We can all relate to this – we know that an idea for a mathematics activity encountered and discussed in a provider-based training session becomes much more real and meaningful when actually applied in practice, working with children. This is a clear example of the way in which provider-based experiences and school-based learning interact and support each other. Teacher knowledge is embedded in the practices of teaching (Hodgen 2011), so that discussing it abstractly is actually quite difficult. However, discussing knowledge in direct relation to school practice, for example in the mentor and pre-service teacher post-lesson discussion, is very productive. Real contexts, events and situations are there, and both mentor and pre-service teacher are able to relate to these readily.

Mentors also need to recognise that their own knowledge, likewise, is situated. In early mentoring experiences, mentors often find it difficult to identify clear examples of mathematics ideas and activities and to give targeted feedback. This is almost certainly due to the fact that much of mentors' knowledge and experience is tacit and intuitive. It is taken for granted and mentors are not accustomed to making it explicit for the benefit of a beginning teacher. Growing in experience as a mentor enables one to develop these skills. The situated nature of teacher learning means that the role of the practice-based mentor in the school context is critical, since much of the beginning teachers' learning and development takes place in school, and the mentor and other colleagues will play a major part in influencing the development of the beginning teacher.

The school environment is a good example of a 'community of practice' (Lave & Wenger 1991). There are certain norms and expectations about how all members of the community behave and act. Certain types of knowledge and skills are required. Teachers and other professionals working in the school environment very quickly learn to take these for granted, but in fact these types of knowledge and skills are complex and specialised. Medical staff in a hospital environment, construction workers on a building site, jazz musicians playing in a band, or navigators on a ship at sea all operate within specific contexts and have to use particular and specialised knowledge. Teachers in the school context also do this. The induction of a beginning teacher into this community of practice is a very significant step. It involves a shift in identity, acceptance into a community, and authentic, meaningful participation in practice. This takes place over time through communication and interaction between the beginning teacher and significant others in the school – in particular, the mentor.

Giving each other permission to feed back

It is important for teachers to be open and honest about their own knowledge and understanding. We acknowledge that, for many teachers, their own learning of mathematics ('mathematical journey') may have been difficult, even threatening, at times. This then can result in a tendency to retreat into defensive behaviour regarding one's own knowledge and understanding. However, this is not productive for the teacher or his/her students. As a responsible educator, I need to give you (my colleague) permission to point out my errors, and offer constructive suggestions for improvement. I need to admit to you areas where I feel insecure, so that we can then start to find ways forward to strengthen my knowledge and confidence. You need to do the same for me.

The components of mathematical learning and their structure

Two qualities, basic ideas and longitudinal coherence, have great significance in terms of a mathematical landscape for learning. Returning to the analogy of Twomey Fosnot and Dolk (2001), the learning landscape is adorned with landmarks that are the 'big ideas' important in the development of students' mathematical understanding. The basic ideas of Ma (1999) are those landmarks of 'big ideas' that students encounter at the beginning of their journey across the landscape. These landmarks and their mathematical connections form the pieces of what Ma terms as 'knowledge packages', the teachers' understanding of the longitudinal process that reveals and nurtures the mathematical landscape in the minds of students. The process by which the student arrives at each landmark is therefore equally as important as the landmark itself, a principle which is acknowledged by both Ma and Twomey Fosnot and Dolk.

Whilst having similarities, these two models also have differences. The landscape of learning is much broader and not only includes the 'big ideas' or key concepts, but also key strategies and models that assist learners in developing their understanding. The knowledge package is more specific in defining the content of the ideas and key ideas required by learners in order to gain an understanding of the main topic. There is also more of a hierarchical structure with centre elements underpinning each other to provide the backbone to which the outside elements are linked. The teaching of these links between the different pieces of knowledge is important. Beginning teachers not only need to know what the big ideas and key concepts are but also how to help students comprehend them and make links between them.

Final remarks

To conclude, we argue that it is vital for mentors to be aware of and help to develop the subject matter knowledge of beginning teachers. Various theoretical frameworks illuminate key ideas in this arena, and some awareness of these ideas will greatly strengthen mentors' understanding and enable them to identify different aspects of mathematics subject knowledge. Regular explicit discussion of mathematics should take place between mentor and beginning teacher, picking up on aspects of teachers' subject knowledge and how they influence children's learning. We contend that this will strengthen not only the beginning teacher's understandings but those of the mentor too. 'Being mathematical' (Watson) or having a 'mathematical sensibility' (Askew) must be traits of a strong mentor who is able to lead by example with a willingness to share and learn, and to do mathematics with beginning teachers, students and more experienced colleagues alike.

References

Askew, M. (2008) 'Mathematical discipline knowledge requirements for prospective primary teachers, and the structure and teaching approaches of programs designed to develop that knowledge', in P. Sullivan and T. Wood (eds) *The International Handbook of Mathematics Teacher Education Volume 1: Knowledge and beliefs in mathematics teaching and teaching development*, Rotterdam: Sense Publishers, pp. 13–36.

Askew, M., Brown, M., Rhodes, V., Johnson, D. and Wiliam, D. (1997) *Effective Teachers of Numeracy: Final Report – report of a study carried out for the Teacher Training Agency 1995–96 by the School of Education, King's College London*, London: King's College.

Ball, D.L. (1989) *Teaching Mathematics for Understanding: what do teachers need to know about the subject matter?*, East Lansing, Michigan: National Center for Research on Teacher Education.

Ball, D.L. (1996) 'Connecting to mathematics as a part of teaching to learn', in D. Schifter (ed.) *What's Happening in Math Class? Volume 2: reconstructing professional identities*, New York: Teachers College Press.

Ball, D.L., Thames, M.H. and Phelps, G. (2008) 'Content knowledge for teaching: what makes it special?' *Journal of Teacher Education*, 59(5), 389–407.

Begle, E.G. (1979) *Critical Variables in Mathematics Education: findings from a survey of the empirical literature*, Washington, DC: Mathematical Association of America and National Council of Teachers of Mathematics.

Byrnes, J.P. and Wasik, B.A. (1991) 'Role of conceptual knowledge in mathematical procedural learning', *Developmental Psychology*, 27(5), 777–86.

Canobi, K.H., Reeve, R.A. and Pattison, P.E. (2003) 'Patterns of knowledge in children's addition', *Developmental Psychology*, 39(3), 521–34.

Chick, H. and Harris, K. (2008) 'Pedagogical content knowledge and the use of examples for teaching ratio', in P.L. Jeffery (ed.) *Australian Association for Research in Education 2007 International Educational Research Conference*, Fremantle, Western Australia: AARE.

Eisenberg, T.A. (1977) 'Begle revisited: teacher knowledge and student achievement in algebra', *Journal for Research in Mathematics Education*, 8(3), 216–22.

Hiebert, J. and Lefevre, P. (1986) 'Conceptual and procedural knowledge in mathematics: an introductory analysis', in J. Hiebert (ed.) *Conceptual and Procedural Knowledge: case of mathematics*, Hillsdale, NJ: Lawrence Erlbaum Associates, pp. 1–7.

Hodgen, J. (2011) 'Knowing and identity: a situated theory of mathematics knowledge in teaching', in T. Rowland and K. Ruthven (eds) *Mathematical Knowledge in Teaching*, London: Springer, pp. 27–42.

Lave, J. and Wenger, E. (1991) *Situated Learning: legitimate peripheral participation*, Cambridge: Cambridge University Press.

Ma, L. (1999) *Knowing and Teaching Elementary Mathematics: teachers' understanding of fundamental mathematics in China and the United States*, Mahwah, NJ: Routledge.

Muir, T. (2008) '"Zero is not a number": teachable moments and their role in effective teaching of numeracy', in M. Goos, R. Brown, and K. Makar (eds) *Navigating Currents and Charting Directions – Proceedings of the 31st Annual Conference of the Mathematics Education Research Group of Australasia*, Adelaide: Mathematics Education Research Group of Australasia, pp. 361–67.

Rittle-Johnson, B. and Alibali, M.W. (1999) 'Conceptual and procedural knowledge of mathematics: does one lead to the other?', *Journal of Educational Psychology*, 91(1), 175–89.

Rittle-Johnson, B., Siegler, R.S. and Alibali, M.W. (2001) 'Developing conceptual understanding and procedural skill in mathematics: an iterative process', *Journal of Educational Psychology*, 93(2), 346–62.

Rowland, T., Thwaites, A., Huckstep, P. and Turner, F. (2009) *Developing Primary Mathematics Teaching: reflecting on practice with the Knowledge Quartet*, London: SAGE.

Schwab, J.J. (1978) 'Education and the structure of the disciplines', in I. Westbury and N.J. Wilkof (eds) *Science, Curriculum and Liberal Education*, Chicago: University of Chicago Press, pp. 229–72.

Shulman, L.S. (1986) 'Those who understand: knowledge growth in teaching', *Educational Researcher*, 15(2), 4–14.

Skemp, R.R. (1976) 'Relational understanding and instrumental understanding', *Mathematics Teaching*, 77, 20–6.

Twomey Fosnot, C. and Dolk, M. (2001) *Young Mathematicians at Work: constructing number sense, addition and subtraction*, Portsmouth, NH: Heinemann.

Watson, A. (2008) 'Developing and deepening mathematical knowledge in teaching: being and knowing', *Nuffield Seminar Series: Mathematical Knowledge in Teaching*. Online. Available HTTP: <http://mkit.maths-ed.org.uk/MKiT5_Watson_distribution-version.pdf> (accessed 3 October 2012).

Wu, H. (1999) 'Basic skills versus conceptual understanding: a bogus dichotomy in mathematics education', *American Educator/American Federation of Teachers*, 23(3), 1–7.

Chapter 2

Developing pedagogical approaches to using digital technologies in mathematics

Rosalyn Hyde and Keith Jones

The evidence suggests that developing effective use of digital technologies in learning and teaching mathematics is a complex matter and that, for pre-service teachers, there is much to be learned in building the appropriate skills to do this. This chapter makes use of an extended model of pedagogical content knowledge that encompasses technological knowledge, skills and understanding as this provides a suitable viewpoint for considering pre-service teachers' learning in this field. It also considers ways in which mentors can support pre-service teachers in the development of an inclusive pedagogy towards digital technologies, and discussion on developing and reflecting on progress in these aspects of teacher learning.

Introduction

The development of digital technologies is characterised by rapid change as computing power increases and device size decreases. The applications of technology are rapidly expanding in every area of life and, increasingly, technology is becoming an essential part of life, including in education. The scope of computer-based hardware available to classroom teachers encompasses devices well-suited to whole-class teaching, such as interactive whiteboards, as well as individual devices for use by students and teachers such as graphing calculators, personal computers, handheld personal digital devices and the like. Alongside the hardware aspect of digital technologies, teachers have access to a huge variety of software and tools for virtual collaboration, social networking, assistive technology, image manipulation, multimedia, data collection, video conferencing and developing learning platforms, offering ever-increasing opportunities for interaction, availability of information and connectivity.

Meanwhile, the inequity in opportunity to use digital technologies between those from different cultures, demographic and socio-economic backgrounds becomes ever wider and more apparent. Young people are often referred to as 'digital natives', implying that they are confident and competent users of digital technologies, but the reality is more complex than this. The use of such technologies in schools is generally planned, collaborative and specific to particular curriculum learning objectives, whereas uses outside school can be characterised as unstructured, exploratory and individual (Grant 2010). This provides a challenge to teachers, both in-service and pre-service, in supporting learners in developing and using their skills appropriately to support learning in mathematics.

Using digital technologies to develop mathematics learning

Pre-service teachers need to start their thinking in this area by understanding the key features of good mathematics teaching with digital technologies. Given the afore-mentioned rapid pace in the development of digital technologies, and the tendency of education to either lag behind or to re-interpret the use of technologies, this is often not the same as their personal use of technology, or their own learning experiences. The National Council of Teachers of Mathematics (NCTM) in America describes effective teachers as those who "maximize the potential of technology to develop students' understanding, stimulate their interest, and increase their proficiency in mathematics" and go on to say that "In a balanced mathematics program, the strategic use of technology enhances mathematics teaching and learning. Teachers must be knowledgeable decision makers in determining when and how their students can use technology most effectively" (NCTM 2008). Good mathematics teaching with technology is therefore about identifying what the benefits are to learners of using technology and what learning the technology makes possible that is not possible without it. Examples of such uses of technology could include using a simulation in the teaching of probability to provide a large set of data in a short period of time so that long-term trends in the data can be observed, using dynamic geometry software to develop an understanding of the constraints of a particular construction through direct manipulation and experimentation, and use of a wiki to develop collaborative illustrated definitions for different polygons or polyhedra.

Pre-service teachers need to be in a position to be able to justify their use of digital technologies in learning and teaching. Such justifications might rightly range from the convenience, efficiency and economy of using activities such as storing test marks electronically or reusing resources, through to encouraging increased engagement, motivation and interest in mathematics, perhaps through the use of video clips, to developing a deeper conceptual understanding of mathematics, through to activities such as the use of graph plotting software to explore the effect of changing the coefficients of a quadratic function.

Another way to consider the key features of good mathematics teaching with technologies is to use the Becta (2009a and 2009b) entitlement documents for primary and secondary school students. These identify six major opportunities for learners to benefit from the use of ICT in mathematics: learning from feedback, observing patterns, seeing connections, developing visual imagery, exploring data and 'teaching' the computer. Table 2.1 provides some examples of each of these opportunities.

Table 2.1 Examples of a learner's entitlement

Opportunity	*Examples*
Learning from feedback	Using a calculator to search for a number that when multiplied by 7 gives the answer 16.
	Using a graph plotter to explore what happens when the coefficients are changed in $y = mx + c$.
Observing patterns	Using an interactive number grid to explore multiples and find primes.
	Using a calculator to explore families of fractions with the same denominator.
Seeing connections	Using dynamic geometry software to resize and reorient shapes and to transform between different shapes with the same number of sides.
	Using graphing or dynamic geometry software to make explicit links between graphical, numerical and algebraic forms of problems.
Developing visual imagery	Using Logo to produce a wallpaper pattern.
	Using an app or web applet to manipulate 3-dimensional objects in space.
Exploring data	Collecting class data and comparing it with a larger national data set online.
	Using an app or web applet to collect data on reaction times and analysing it.
'Teaching' the computer	Designing and testing procedures in Logo.
	Using the formula function of a spreadsheet to solve a problem.

The phrases in the Becta entitlement documents provide a powerful framework for pre-service teachers to reflect on, and use, to identify effective ways of using digital technologies in their lessons and to seek out examples for the areas they have not yet experienced or used in their teaching. Such identification might be undertaken by auditing a school scheme of work, through lesson observation, or by using the list to identify opportunities in the pre-service teacher's own lesson planning, and therefore providing a good starting point for considering pedagogic uses of digital technologies.

Interactive learning with digital technologies

One of the features of using digital technologies in learning and teaching mathematics that is most often identified as a positive benefit of its use is that of 'interactivity'. This is a word that is frequently used to describe effective teaching, although it is often not well defined. Kennewell *et al.* (2007) found in their research that digital technologies can be used to support interactive teaching in three main ways:

1 as the *object* of interaction
2 as a *participant* in interaction
3 as a *tool* for interaction

Some illustrative examples of these are provided below.

Learners might discuss a projected digital image such as the one in Figure 2.1, where the interactivity is provided by a discussion *about* the example of digital technology.

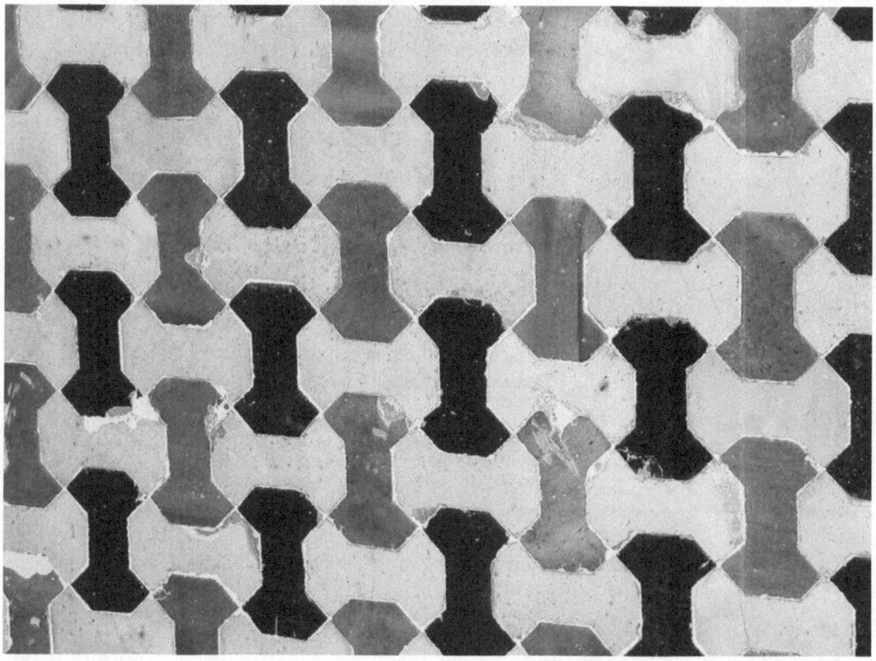

Figure 2.1 Tiling pattern

The second type of interactivity, where digital technology is a *participant*, might be provided by an activity such as the one in Figure 2.2, where learners receive instant feedback each time they enter a command.

The final type of interaction, where digital technology is a tool for *interaction*, is the most powerful in supporting teaching and learning because here the use of digital technologies directly supports conceptual learning and there is the opportunity for skilled conversation between learners and the teacher, and between

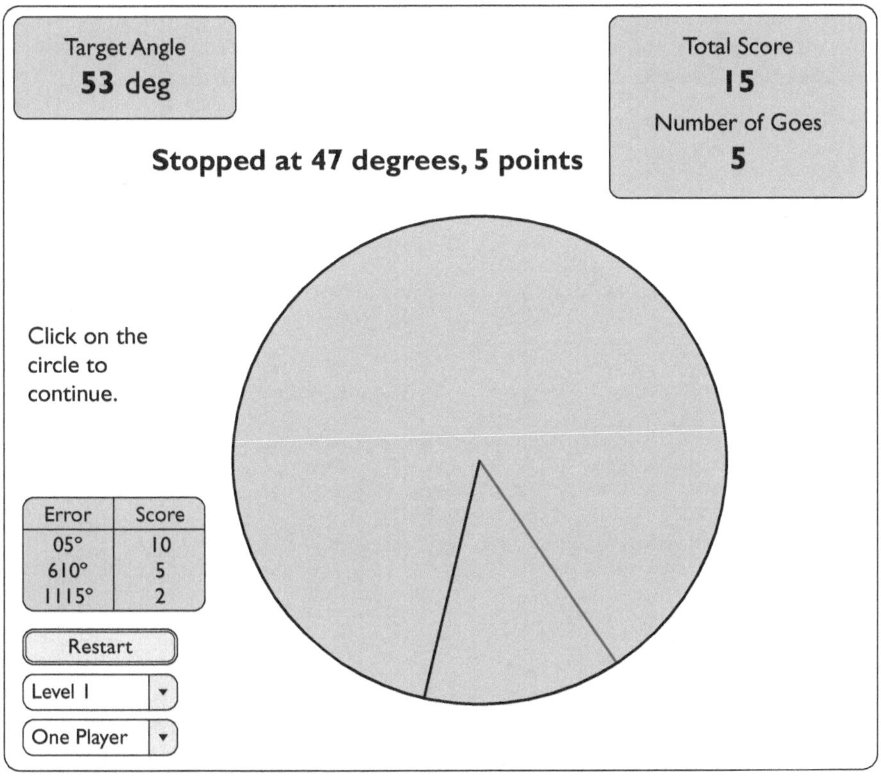

Figure 2.2 Estimating angles (NRICH). The NRICH website http://nrich.maths.org publishes free mathematics resources designed to challenge, engage and develop the mathematical thinking of students aged 5 to 19. NRICH also offers support for teachers by publishing Teachers' Resources for use in the classroom.

learners themselves to develop shared understanding. In the example in Figure 2.3, multiple representations, the use of sliders to allow learners to control the variables and the use of a shared projected image allow learners to explore misconceptions about distance-time graphs and to make connections between these and the gradient–intercept form of a straight line equation.

A place to start when considering interactivity might be for the pre-service teacher to notice who it is who is using and controlling the digital technology in the classroom. When the technology remains in the hands of the teacher in a somewhat passive manner, the question is raised as to the place of interactivity in the observed learning. Discussing the difference between *technical* interactivity and *pedagogic* interactivity also provides a way into thinking about the mathematics learning taking place and how the learning has been facilitated.

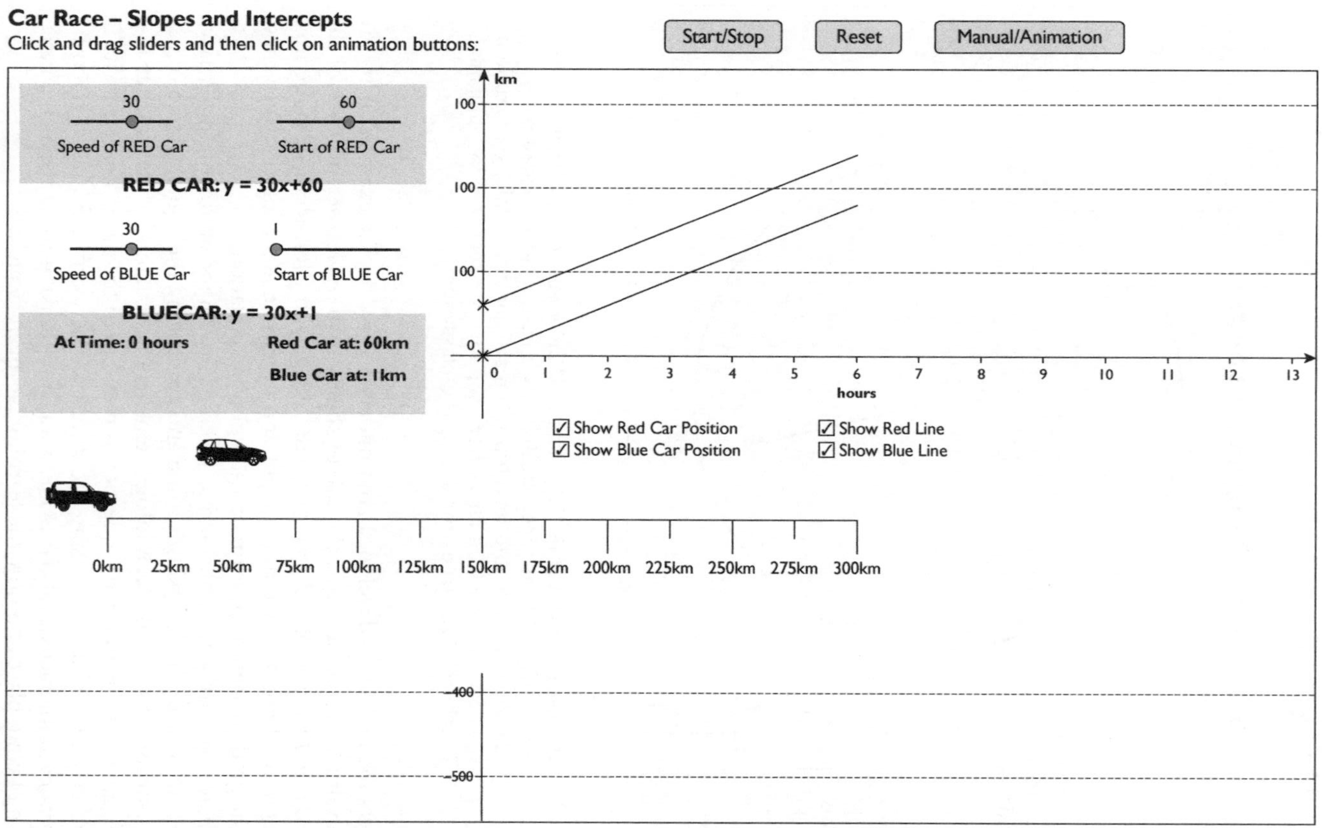

Figure 2.3 Car race – Slopes and Intercepts from Technology for Teaching 'Thinking Mathematics' (http://www.mathcasts.org/mtwiki). Interactivity created in Geogebra (www.geogebra.com)

- a collaborative classroom where learners work together and discuss their mathematical experiences using digital technologies as appropriate to support their work and where teachers and learners share a sense of responsibility for learning mathematics;
- a wide variety of teacher and learner interactions are in evidence to explore and expand learners' understanding around the mathematical concepts and ideas stimulated by the digital technologies-generated problems;
- learners are engaged and actively involved in the exploration and making sense of mathematics (using digital technologies) through meaningful activities that make connections across all areas of mathematics;
- computer projection displays are used to provide instant feedback to increase learners' cognitive engagement, not only for purposes of demonstration and assessment.

Figure 2.4 Features of interactive use of digital technologies in mathematics classrooms (NCETM 2010: 6)

Figure 2.4 provides a list of features of interactive use of digital technologies in mathematics teaching. This might provide a useful framework for mentors to use when observing pre-service teachers using digital technologies in their teaching.

Embedded practice with digital technologies

The Joint Mathematical Council (JMC) report on digital technologies in mathematics education (2011) promotes what is referred to as 'embedded practice' where students use digital technologies to "find out about, learn, apply and communicate aspects of mathematics" and the "authentic use" of digital technologies in "computer algebra, modelling, simulation, statistics, data-capture and 3D geometry" (p. 16). Whilst the report is critical of a perceived lack of progress in the use of digital technologies in these two areas, these are goals for digital technology use the development of which all who work with pre-service teachers should be supporting in the next generation of teachers.

A straightforward use of digital technology for communicating mathematics that a pre-service teacher might undertake with a class, with access to suitable resources or as an out-of-school task, is for learners to produce a report of some kind about a mathematics project or open-ended task. The report could be in the form of a video, a poster, a written report with supporting images, or produced using presentation software. An approach where a teaching placement is seen as a place to try things out in a 'safe' and supported environment where mistakes are followed by analysis and reflection gives pre-service teachers confidence. Once confident, they are then able to try open activities and extended activities lasting over several lessons, including activities where digital technologies might be used to find out about, learn and apply mathematics. One example might be the use of the Internet and a spreadsheet to explore hypotheses about the nutritional content of different foods, an activity that demands use of various elements of number, measurement and data handling.

Comments such as those in the JMC report quoted previously point to a need for school mentors to support pre-service teachers in critically exploring observed practice in their teaching placement school where aspirations do not always match expectations for myriad reasons, some of which will be explored later. Embedding the use of digital technologies into practice and developing authentic uses of tools are demanding for pre-service teachers but are aspirational for them as they develop further as more experienced teachers. Both more importantly and more manageably, pre-service teachers can be supported to develop good practice in making interactive uses of digital technologies.

What remains central to effective use of digital technologies in learning and teaching mathematics is that the use of the technology benefits the learning in some way. This might be through providing engagement and motivation for learning, but it should also be through providing opportunities for learners to interact and experience mathematics in new and more powerful ways, leading to new learning, that are not possible without the technology.

Uses of digital technologies in mathematics

Interactive whiteboards began appearing in UK classrooms in the early 2000s. Since then, they have become one of the most widely used tools in UK classrooms. As such, pre-service teachers need to develop some basic skills in their use very early on in their period of initial teacher education. This is best achieved by arranging for them to explore the software and hardware in an empty classroom so they are familiar with the particular system used in their placement school or college. Also valuable is observation of more experienced teachers who make creative uses of a range of functions available on the interactive whiteboard such as hide/reveal, using the drawing and mathematical tools, importing images, highlighting, colouring, drag and drop and so on. A practice run-through of a lesson in an empty classroom can also provide opportunities for the pre-service teacher to build confidence with the tools they need to use, rehearse their lesson and check their work to see if there are any issues with things such as image and text sizing and colouring.

Pre-service teachers might usefully consider alternative ways of using an interactive whiteboard in the classroom other than whole class teaching or teacher presentation. Such boards often hold learners' attention, and involving them through direct interaction and control of the board can both increase engagement in lessons as well as develop more learner-led collaborative thinking in mathematics lessons.

Another tool many pre-service teachers are initially unfamiliar with is dynamic geometry software. Such software is more widely used in secondary schools than primary schools, although there are junior versions of the most commonly used packages. There are many applications in the classroom across the mathematics curriculum. It provides, for example, valuable hands-on exploration for primary-aged learners to explore the properties of polygons as well as for much older learners to explore functions and their derivatives.

Web sites such as www.censusatschool.org.uk provide learners with the opportunity to work with large sets of real data. Amongst the many resources available on the site there is an interactive histogram where it is possible to change the size of the groups along the horizontal axis of a histogram and see the effect on the diagram. Data loggers provide opportunities for learners to collect data for themselves before analysing it. There are many more examples that could be added here to demonstrate effective use of digital technologies in the mathematics classroom. Interested readers might usefully refer to Oldknow, Taylor and Tetlow (2011) for further examples.

Developing technical skills

Pre-service teachers need both a technical understanding of the digital technologies they will be using in the classroom as well as understanding about pedagogy, both for teaching mathematics and teaching mathematics with technology. This leads us in two directions at this point: first, to discuss the need for technical skills with technology and second, to explore a framework for understanding the relationships between technology, pedagogy and subject knowledge.

Most accrediting bodies for initial teacher education have systems in place for auditing, tracking and monitoring the skills pre-service teachers have in using digital technologies. These are often self-reporting and should track skill-confidence levels as well as classroom competence. Many of those starting teacher preparation programmes have good personal skills in using some technological tools through their own studies, daily life or previous careers. However, they may not have had any experience of using graphing programmes (such as Autograph), dynamic geometry software (such as Geogebra, Cabri or The Geometer's Sketchpad) and Logo. Such technical skills can be provided through specialist-taught sessions provided though a training provider or institution, through self study, or through the pre-service teacher joining with a class of students whilst on school placement. Whatever the means, the opportunity to explore software and 'play' as well as having a purpose for learning to use such software serves an important role in the learning process.

Technical skills alone, however, are not sufficient to ensure good teaching with digital technologies. All of us have met 'techies' who have poor communication skills and have seen examples of technology used where a non-digital tool, such as pen and paper, would have been just as effective and, in some cases, better. Some technical skill is necessary, but in order to use technologies effectively for learning this is only part of the picture. For example, Mercer, Hennessy and

Warwick (2010) are clear that, in the case of interactive whiteboards, "the effective use of the IWB as an educational tool is not inherent in the hardware, software or even the materials it displays. It is predicated upon the teachers' practical understanding of how to engage students and to help them learn" (p. 207). Developing the effective use of digital technologies is therefore rooted in understanding pedagogy.

Types of knowledge needed to teach with technology

Mishra and Koehler's (2006) framework articulating technological pedagogical content knowledge, developed from the work of Lee Shulman (1986) about teacher knowledge, provides us with one way of exploring aspects of necessary learning for pre-service teachers in learning to make effective use of digital technologies in learning and teaching.

Shulman (1986) is well known for his work in developing thinking in the field of the knowledge needed for teaching. His view is that teachers, in addition to content knowledge relevant to the subject being taught and general pedagogical knowledge about teaching, also need a type of knowledge that combines both and is specific to subject teaching, which he called pedagogical content knowledge.

The two types of knowledge, content (C) and pedagogical (P), can be represented using a Venn diagram (see Figure 2.5) with the new type of knowledge identified by Shulman, pedagogical content knowledge (PC), being formed in the overlap of the first two types of knowledge.

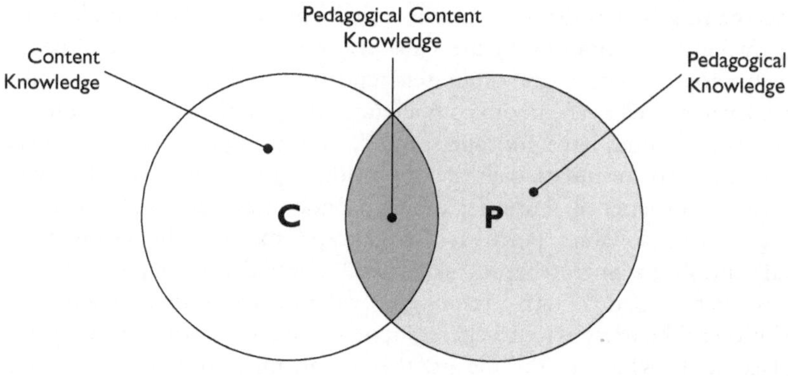

Figure 2.5 Pedagogical content knowledge

Mishra and Koehler (2006) developed Shulman's work further by including a third circle labelled technology knowledge (T) and overlapping this with the diagram in Figure 2.5 to form new categories of knowledge called technological pedagogical knowledge (TPK), technological content knowledge (TCK) and, where all three circles overlap, technological pedagogical content knowledge (TPCK). These are illustrated visually in Figure 2.6.

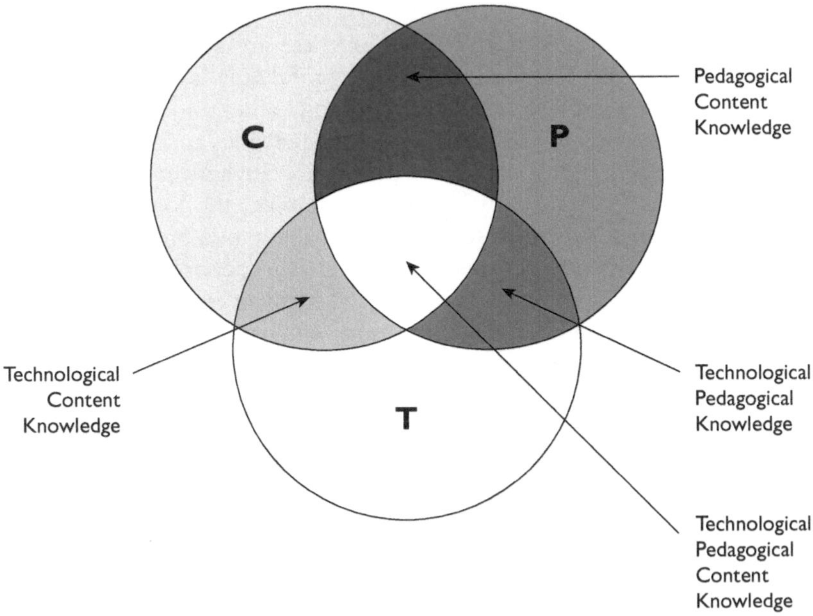

Figure 2.6 Adapted from Mishra and Koehler's (2006) framework for Technological Pedagogical Content Knowledge

In terms of digital technologies, technology knowledge includes the skills required to operate hardware such as computers and digital projectors, to use 'standard' software tools such as word processors, web browsers and spreadsheets, and to manage documents, programs and files. Technological content knowledge is about having an understanding of how technology and content, in this case mathematics, are related to one another. The affordances of digital technologies allow for multiple representations of mathematics to be explored concurrently and for changes in one form, for example a table of data, to automatically change into another form, such as a graph. A commonly used mathematical example of this type of knowledge is that of dynamic geometry software, which allows learners to construct geometrical proof through 'playing' with the software and allowing the mathematical knowledge to be constructed in a way that is not possible without the software. Other examples could include understanding how to access the Internet, how a spreadsheet allows students to learn statistics with real data and how using Logo and floor turtles allows students a tangible experience of angle as a measure of turn.

Technological pedagogical knowledge concerns the generic use of technology in teaching. This relates to understanding how different tools and technologies can be used and how to select and use the appropriate tool in pedagogically appropriate ways in the classroom. It also includes knowing how to use tools for administrative purposes in teaching and how to use generic tools such as discussion boards and chat rooms that may be part of a whole school 'virtual learning environment' (VLE).

Clearly, pre-service teachers need skills in all these areas in order to teach effectively with technology. The need for good technical skills is discussed earlier in this chapter. Making explicit how the technology and content interact by identifying how technology has changed school mathematics and how it might change it further can be a useful activity for pre-service teachers to undertake to develop technological content knowledge. For example, the log button and buttons for trigonometric functions supersede the paper tables previous generations used in their mathematics lessons. Opportunities to develop technological pedagogical knowledge can be effectively provided by mentors through focused observation across a school or college, support with hands-on use of digital technologies by pre-service teachers both in and out of lessons, and through the use of reflective approaches to evaluation.

Mishra and Koehler argued that teachers need more than these previously described skills to make effective use of technology and they therefore developed the notion of technological pedagogical content knowledge (TPCK) as being at the place of intersection of technology, pedagogy and content knowledge. It is described as "the basis of all good teaching with technology" and explained by saying "developing good content requires a thoughtful interweaving of all three key sources of knowledge: technology, pedagogy and content" (2006: 1039). This is clearly a sophisticated process that takes time (perhaps a lifetime!) to develop. Our aim, in working with pre-service teachers, is to develop teachers who have started their learning journey in developing learning environments for students that utilise these ways of connecting content knowledge with pedagogical and technological knowledge.

It is not necessarily true that these skills need to be developed in a pre-determined or hierarchical order. This chapter argues earlier that some degree of technical skill is necessary but that technical skill alone is insufficient for effective teaching and learning. Pre-service teachers are likely to be able to demonstrate, with a specific example of technology use, that they can effectively connect content knowledge, technological knowledge and pedagogical knowledge. They then need to move on to use other technologies in other parts of the curriculum in effective ways. This process is important, particularly because, as Mishra and Koehler suggest, even for experienced teachers, incorporating an unfamiliar technology forces the teacher to reconsider what they are doing and, as Mishra and Kohler put it, reconstruct "the dynamic equilibrium among all three elements" (2006: 1030).

Practicalities

One of the challenges facing pre-service teachers in developing use of technology in their teaching is in arranging for learners to have opportunities to work in a computer suite/laboratory. In many schools and colleges, the use of a computer suite needs to be booked well in advance of the intended lesson and it can be difficult for pre-service teachers to anticipate what they will be teaching sufficiently far in advance. Some support from a more experienced teacher in identifying curriculum opportunities in advance can allow them to plan more effectively.

Teaching in a computer suite can be very different to teaching in a more traditional mathematics classroom and the necessary technical pedagogical knowledge required can seem daunting to the pre-service teacher. Team teaching and shared planning with another teacher (or pre-service teacher) is one way of supporting pre-service teacher learning in this area whilst providing a 'safer' environment in which to practice. An alternative is for the pre-service teacher to take only part of the class into a computer suite whilst the class teacher teaches the remainder in the usual teaching room.

Pre-service teachers may need to be prompted to consider the implications the layout of the computer suite will have on classroom management. It may be, for example, that there is no whole-class projector and screen in the computer room, or that some students won't be able to see clearly from where they are sitting. Careful attention needs to be paid to how instructions are given; for example, at what point students are instructed to start using the computer, how additional printed material might be used to support students, and whether students should work individually or in pairs. Strategies to monitor students' work and their progress might need to be adapted as well as consideration given at the planning stages as to how students will record their learning. Similar issues need to be considered when using a class set of laptop or tablet computers.

Of course, effective use of digital technologies takes place in environments and contexts other than computer suites. In many primary classrooms learners work in groups and one of those groups might have using some software on the interactive whiteboard as their activity. In this way, the large touch-sensitive screen can provide a powerful environment for collaborative learning where learners are able to 'interact with the mathematics' in a direct way. In other classrooms, the presence of four or five computers provides an opportunity for a 'carousel' approach, where learners rotate round a series of different activities, one of which could involve using the computers, during the course of a lesson or series of lessons.

A growing number of schools have provision where all learners, or all learners in a certain group, have a personal laptop or tablet computer that they take with them from class to class. This level of availability of hardware changes the dynamics of use of technology in the classroom and, again, the implications of this type of resourcing need thinking through by pre-service teachers, particularly in terms of integrating technology into lessons and using laptops or tablets in a responsive way. Further discussion of tablets and other mobile devices is undertaken later in this chapter.

Affordances and constraints for pre-service teachers

Learning to use digital technologies in teaching is not always straightforward. Evidence from Ofsted school inspections in England continues recent criticism about use of Information and Communication Technology (ICT) in schools to teach mathematics claiming that "the potential of ICT to enhance learning in mathematics continues to be underdeveloped. The principal uses of ICT in the lessons observed were for presentation of material by teachers in whole-class teaching and by pupils for revision, practice and homework" (Ofsted 2012: 28).

The JMC digital technologies report (2011) says that the "barriers to the more pupil-focussed use of digital technologies" include:

- inadequate guidance concerning the use of technological tools within both statutory and non-statutory curriculum documentation;
- current assessment practices, which do not allow the use of digital technologies, particularly within high stakes assessments;
- a perception that digital technologies are an add-on to doing and learning Mathematics.

(p. 6)

These reports from Ofsted and JMC illustrate the situation faced by many mentors and pre-service teachers in school and college environments. Whilst the first two reasons listed above from the JMC report are outside the direct control or influence of classroom teachers, it is possible for mentors and pre-service teachers to make effective pedagogical use, as described above, of those resources they do have available.

It is worth noting that Hammond *et al.* (2009) found that those pre-service teachers who were very good users of digital technologies were also very good teachers. In making such a broad statement it is important to note that 'very good users' of digital technologies were considered to be those pre-service teachers who made frequent, varied use of digital technologies and had the ability to focus on the use of digital technologies to support subject teaching. 'Very good teachers' were seen to be those pre-service teachers who were attaining highly against nationally set expectations for qualifying as a teacher. The findings of Hammond *et al.* support an argument that developing the skills to teach well with digital technologies is complex and is closely linked to the pre-service teacher developing good teaching skills in other areas of their practice. However, it does not suggest that weaker pre-service teachers aren't capable of making good use of technology in their teaching.

Hammond *et al.*'s study goes on to suggest that the use of digital technologies by the new teachers with whom they worked can be attributed to several factors:

- access to ICT;
- a perception that it helped them meet their aims as a teacher and provided opportunities not otherwise available;
- an inclination to use it based on previous personal and academic use;

- resourcefulness in overcoming constraints and identifying opportunities;
- pre-service training involving the direct modelling from tutors at university and mentors in school alongside peer support;
- general support and encouragement for using ICT in school.

(p. 104)

These findings demonstrate that, whilst access to resources is important, there are other factors that make strong contributions to the development of effective pedagogical use of digital technologies. Noticeable is the implicit importance of the mentor in the development of the pre-service teacher's skills and the positive personal qualities of the pre-service teacher.

Affordances and constraints for mentors

For some mentors, mentoring a pre-service teacher in the effective use of digital technologies in learning and teaching may feel quite challenging, particularly if the mentor feels under pressure 'to be the expert'. This might be particularly true where the mentor feels that they are still working on their own technology skills while mentoring a pre-service teacher who has a very high level of technical skill.

There are a number of possible ways of working with this scenario: this has at its root the relationship between mentor and mentee. Mutton, Mills and McNicholl (2006) are clear that it is not a necessary condition that a mentor be an expert in the use of digital technologies before they can be an effective mentor in this aspect of pre-service teacher learning. They suggest a number of ways forward using different models of collaboration where mentor and pre-service teacher work together and that allow the mentor to share their pedagogical expertise. Such working together might be facilitated by a collaborative approach where both participants draw on the different types of knowledge and skills of the other for joint professional learning through joint planning and discussion. Alternatively, a more collegial model of mentoring might be more appropriate, where both participants both recognise the contributions they make as experts and recognise their role as learners. Such approaches are discussed further in Chapter 8.

Working in such ways with a pre-service teacher who is more capable with digital technologies than their mentor can offer some affordances if both parties can adapt to working in one of the ways suggested above. For the mentor there is both the opportunity to improve their own skills with digital technologies by learning from the pre-service teacher but also to reflect on the adaptation of their own pedagogies to the use of the technology. For the pre-service teachers there is value in being able to use an area of strength, in this case in using digital technologies, in learning and teaching.

The future

One thing is certain: digital technologies will continue to develop at a rapid rate and in ways that we are yet to see and experience. Such developments are having huge impacts on our social, recreational, work and education lives.

The 2012 *Horizon Report* (Johnson, Adams & Cummins 2012) envisages the six emerging technologies or practices that are likely to enter the mainstream of schools within some given time scales. These are:

- Within one year or less
  - mobile devices and apps
  - tablet computing
- Within two to three years
  - game-based learning
  - personal learning environments
- Within four to five years
  - augmented realities
  - natural user interfaces.

These offer huge opportunity alongside massive challenge to educators. The enormous explosion in the availability and variety of apps available for mobile devices and tablets provides a lot of opportunity for teaching and learning mathematics. At the time of writing, there are graph plotting apps available as well as versions of Logo and simulations, but no versions yet available of dynamic geometry software. There are also innumerable tools that might find uses in the mathematics curriculum such as those for measurement, calculation, conversion, equation solving, etc. There are also new ways of interpreting teaching and learning through developing and sharing material and through collaborative working both by teachers and learners. Instant access to tools and materials on such a scale changes some of the notions of what learning is and how and where it happens.

Final remarks

Thinking about the use of digital technologies in the classroom can force the pre-service teacher and their mentor to face up to some uncomfortable questions. For both, there is the individual challenge of the technical demands of unfamiliar and new technologies. For pre-service teachers there is the demand of simultaneously developing all the kinds of knowledge identified by Mishra and Koehler (2006). For some pre-service teachers, using digital technologies effectively in teaching and learning challenges their beliefs about effective learning and teaching mathematics, and painful adaptations may have to be made as part of the learning process.

Digital technologies are here to stay and will continue to develop. Educators need to be using these technologies for learning, but it remains central that uses of digital technology in teaching mathematics need to be those that benefit the learner's understanding in some way. As teachers, and mentors working with pre-service teachers, we need to be supporting emerging generations of teachers in developing the skills they need to both work in such a world and to work with young people who live and will work in such a world.

References

Becta (2009a) *Primary Mathematics with ICT: A Pupil's Entitlement to ICT in Secondary Mathematics*, Coventry: British Educational and Communications Technology Agency.

Becta (2009b) *Secondary Mathematics with ICT: A Pupil's Entitlement to ICT in Secondary Mathematics*, Coventry: British Educational and Communications Technology Agency.

Grant, L. (2010) *Developing the Home–School Relationship using Digital Technologies*, Bristol: Futurelab.

Hammond, M., Crosson, S., Fragkouli, E., Ingram, J., Johnston-Wilder, P., Johnston-Wilder, S., Kingston, Y., Pope, M. and Wray, D. (2009) 'Why do some student teachers make very good use of ICT? An exploratory case study', *Technology, Pedagogy and Education*, 18(1), 59–73.

Johnson, L., Adams, S., and Cummins, M. (2012) *NMC Horizon Report: 2012 K–12 Edition*, Austin, Texas: The New Media Consortium.

Joint Mathematical Council (2011) *Digital Technologies and Mathematics Education*, London: Joint Mathematical Council.

Kennewell, S., Beauchamp, G., Jones, S., Meiring, L., Morgan, A., Norman, N., Parkinson, J., Tanner, H. and Thomas, D.G. (2007) *The Use of ICT to Improve Learning and Attainment Through Interactive Teaching: Final Report to ESRC*, (expanded), Swansea: Swansea Metropolitan University.

Mercer, M., Hennessey, S. and Warwick, P. (2010) 'Using interactive whiteboards to orchestrate classroom dialogue', *Technology, Pedagogy and Education*, 19(2), 195–209.

Mishra, P. and Koehler, M. (2006) 'Technological pedagogical content knowledge: A framework for teacher knowledge', *Teachers College Record*, 108(6), 1017–54.

Mutton, T., Mills, G. and McNicholl, J. (2006) 'Mentor skills in a new context: working with trainee teachers to develop the use of information and communications technology in their subject teaching', *Technology, Pedagogy and Education*, 15(3), 337–52.

National Centre for Excellence in the Teaching of Mathematics (NCETM), (2010) *Mathematics and Digital Technologies – New Beginnings: A Report*, London: NCETM.

National Council of Teachers of Mathematics (NCTM) (2008) *The Role of Technology in the Teaching and Learning of Mathematics*. Online. Available HTTP: <http://www.nctm.org/about/content.aspx?id=14233> (accessed 3 October 2012).

NRICH, *Estimating Angles*. Online. Available HTTP: <http://nrich.maths.org/1235> (accessed 21 November 2012).

Office for Standards in Education (Ofsted) (2012) *Mathematics: Made to Measure*, London: Ofsted.

Oldknow, A., Taylor, R. and Tetlow, L. (2011) *Teaching Mathematics Using ICT*, 3rd edn, London: Continuum.

Shulman, L. (1986) 'Those who understand: knowledge growth in teaching', *Educational Researcher*, 15, 4–14.

Chapter 3

Approaches to learning mathematics

Taro Fujita and Rosalyn Hyde

The way teachers teach mathematics depends strongly on their own personal view of what mathematics is and how it is learnt. Exploring different views of the nature of mathematics, what mathematics is, and what it means to learn and teach mathematics is very important for pre-service teachers. We discuss the importance of understanding the 'big picture' and 'big ideas' in mathematics before taking the viewpoint of mathematics as the 'science of patterns'. We also show that Wittmann's (1995) idea of a 'Substantial Learning Environment' is useful for developing rich tasks for learners of mathematics.

Introduction

Mathematics is a rich and complex subject that has fascinated and intrigued people for centuries. It is also an area of study that many find both difficult and frustrating and about the teaching of which everyone seems to have an opinion. Beliefs about the nature of the subject and how it should be taught are central to the pre-service teacher's developing notions of themselves as a teacher of mathematics and how they work with learners in the classroom. Communicating a developing understanding of mathematics as a connected body of knowledge with over-arching themes is important for pre-service teachers' own learning and that of their students. Equally, and connected to these ideas, pre-service teachers need to appreciate what makes a good mathematics task for learners and to develop classroom environments that stimulate mathematics learning through the use of rich tasks.

Beliefs about mathematics and teaching

There is considerable evidence that the beliefs teachers hold about the nature of mathematics have a significant impact on their classroom practice. Studies also suggest that when new curricula are introduced, their success is often dependent on teachers' beliefs about mathematics and teaching. When teachers can fit the expectations of new curricula and reform into their existing beliefs about mathematics and its teaching, they will successfully implement the new curriculum.

Where there is some mismatch, they will either adapt the new curriculum to fit their beliefs and therefore only partially implement it, or they will not implement it to any great extent. In a similar way, pre-service teachers may make adaptations to the intended curriculum as a result of their beliefs about mathematics and its teaching. We also know that the biggest influences on how pre-service teachers perceive teaching and learning are their own experiences as learners. Such beliefs can be resistant to change and can also be internally inconsistent.

It is therefore important that pre-service teachers and mentors are able to identify their own beliefs about mathematics and about teaching, and recognise the impact these have on their practice. Ernest (1989) suggests that the three key belief components are the teacher's:

- view or conception of the nature of mathematics;
- model or view of the nature of mathematics teaching;
- model or view of the process of learning mathematics.

The nature of mathematics

Pre-service teachers come to a career in teaching from a range of backgrounds and at different points in their working lives. In our experience, prospective teachers have generally been successful at school themselves. Those intending to teach at secondary level, requiring a high level of expertise in mathematics, starting their teaching careers in their early twenties, have often been very successful academically in mathematics or a closely related discipline. Those entering teaching later in their working lives may also have been successful in other fields, and often have work experience using mathematics in a range of fields such as engineering, business and finance. Some pre-service teachers may also be required by their accrediting institution to undertake further study in mathematics prior to starting their initial teacher education. For those intending to teach primary age learners, statistics suggest that very few have a degree in mathematics or a mathematics-related subject. For some, whilst they meet the minimum level of competency in mathematics in order to enter the teaching profession, mathematics is a subject where they lack confidence in their own ability and in their ability to teach it. For others, their experience of learning mathematics has been more pleasurable and successful. Such a range of backgrounds for those beginning their initial teacher education is likely to lead to individuals holding a broad range of views on the nature of mathematics.

Some pre-service teachers have not given a great deal of thought to the beliefs they hold about mathematics, or to the possibility that others may view mathematics differently to themselves. In fact, the question 'What is mathematics?' is a difficult one to answer. If you pose this question to ordinary people then you might hear from them a simple answer such as 'calculation' or 'times table' Pre-service teachers in primary and secondary schools may also suggest various key words such as 'numbers', 'calculation', 'algebra', 'shapes', 'measurement', 'problem solving', 'logic', 'useful', 'interesting', 'confusing', 'terrifying', and so on, based on their learning experiences. Prospective teachers may also express a perception of mathematics as a 'service' subject, where mathematical techniques are learnt in order to be applied to other disciplines, such as engineering. Others might see mathematics as a set of rules to follow and measure success in terms of successful completion of short, closed problems all of a similar nature. It is also not unusual for some to view number and algebra as the core topics in mathematics, with statistics, in particular, being seen as the 'poor relation'. For some, mathematics has a strong underpinning about pattern and relationships. Others might also see it as being a tool for exploring and interpreting the 'real world' or as having a communicative feature in being like a 'language'. These are views about what might be called 'school mathematics'. One might expect professional mathematicians to have their own philosophies about the nature of mathematics. For example, on reading the introductory chapter in a book such as *Philosophy of Mathematics: An Anthology* (Jacquette 2002), one is likely to be overwhelmed by the considerations that need to be addressed in order to answer the question 'What is mathematics?'

The views expressed by teachers and pre-service teachers give us some indications of their underlying beliefs about the nature of mathematics. Aspects of the views expressed previously can be seen as part of the evidence available of pre-service teachers holding different philosophies about the nature of mathematics.

Other evidence of teachers' views of mathematics can be, of course, inferred from their actual classroom practice. Ernest (1989) provides us with one way of categorising the philosophies underlying different beliefs by suggesting that there are three frequently held philosophies of mathematics amongst teachers of the subject. He describes *instrumentalists* as those who see mathematics as an accumulation of facts, rules and skills to be used for a particular purpose. He also describes a '*Platonist*' view of mathematics where mathematics is seen as a "static but unified body of certain knowledge" (1989: 250) and finally a *problem solving* view where mathematics is seen as a "dynamic, continually expanding field of human creation and inventions, a cultural product" (1989: 250).

Beliefs about teaching and learning

Ernest further suggests three possible beliefs about the role of the teacher: instructor, explainer and facilitator. He describes the instructor role as having skills mastery and correct performance as the intended outcome, the explainer role as intending to lead to "conceptual understanding with unified knowledge" (p. 251) and the facilitator role as leading to confident problem posing and solving. Ernest also suggests four simplified models for learning: (1) compliant behaviour and mastery of skills, (2) reception of knowledge, (3) active construction of understanding, and (4) the exploration and autonomous pursuit of the learners' own interests.

Exploring the views prospective teachers have about teaching and learning when they attend interview through questions about their own learning, those who have been role models and mentors to them, their observations of mathematics being taught, and exploration of how they might work with learners themselves, can provide important early insight into their beliefs in these areas. Again, evidence for different beliefs about teaching and learning come both from what is explicitly expressed and what is implicitly inferred in the practice of teachers, but the latter is difficult to observe early on in an individual's initial teacher education.

Clearly, one might think it reasonable that certain beliefs about the nature of mathematics might predispose one to have particular views on the teaching and learning of mathematics. Whilst this is so, it is not always the case. There are teachers who hold self-contradictory beliefs and those whose observed practice does not always match well with their expressed views on mathematics, teaching and learning. For example, pre-service teachers who hold a somewhat relational understanding of, for example, the quadratic formula, might teach this as a rule to memorise because it might be quicker and easier to 'teach' to students as a rule to apply rather than spending more time to explain why this formula can be deduced from $ax^2 + bx + c = 0$.

It is important that mentors recognise that mathematics teachers, both pre-service and experienced, have a range of beliefs about mathematics teaching and learning. One way to facilitate this might be for mentors to work with pre-service teachers in identifying evidence of different views on mathematics, teaching and learning implicit in their own lessons and those of others. This becomes a valuable activity for pre-service teachers given that learners are often very adept at identifying teachers' views and preferences in the classroom. Understanding something of what views are being perceived by learners is therefore helpful.

When mentors and pre-service teachers hold differing belief systems, or when mentors expect pre-service teachers to teach in a way that doesn't fit with their belief system, difficulties may arise. Whilst beliefs about the nature of mathematics and its teaching and learning are remarkably resistant to change, there is evidence to suggest such changes are most likely to occur early on in a teacher's career. One influencing factor supporting change is a pre-service teacher's level of consciousness of their own beliefs and their ability to reflect on their own practice. Reflection supported by a perceptive mentor can allow for recognition and the development of a closer match between belief and practice.

Instead of asking 'What is mathematics?' it might be useful to investigate another question: 'What view of mathematics will be useful to design mathematically interesting and rich tasks in schools?' as a starting point for reflection. Consider the view 'mathematics is a science of patterns'. This view is expressed by several mathematicians and physicists (Hardy 1940; Sawyer 1955: 12; Feynman 1968: 14). More recently, Devlin states that "What the mathematician does is examine abstract 'patterns': numerical patterns, patterns of shape, patterns of motion, patterns of behavior, voting patterns in a population, patterns of repeating chance events, and so on" (Devlin 1998: 3). We would like to stress that this is not the *only* useful view of the nature of mathematics. Rather, starting from this view might provide opportunities to enrich pre-service teachers' understanding of various views of the nature of mathematics and of teaching and learning. In what follows we discuss and show how this simple view of mathematics can be a powerful vehicle to challenge or reconsider common views and beliefs about mathematics and to design mathematically rich tasks in primary and secondary schools.

The big picture in mathematics

As we have seen, Ernest discusses the three views of mathematics, '*instrumentalist*', '*Platonist*' and '*problem solving*'. We take the final view, mathematics as '*problem solving*', with its emphasis on connections in mathematics, as the most appropriate for school mathematics.

One way to explore this view of mathematics is by drawing a concept map. For example, a mentor and a pre-service teacher might sit together and try to draw a concept map of mathematics. While this provides us with a good starting point for discussion, our experience is that, whilst people will be satisfied when they see how physically 'big' the map is, and how many lines are used to connect concepts, it is still questionable whether the map conceptualises mathematics from a (conceptually) big point of view. The examples in Figure 3.1a and b are concept maps drawn by pre-service teachers in primary schools.

At first glance, these concept maps list various elements of mathematics and show that mathematics is conceptualised as a big picture. However, we also notice that whereas some elements are certainly important, there are few connections between elements of, for example, numbers and shapes. Considering the impor-

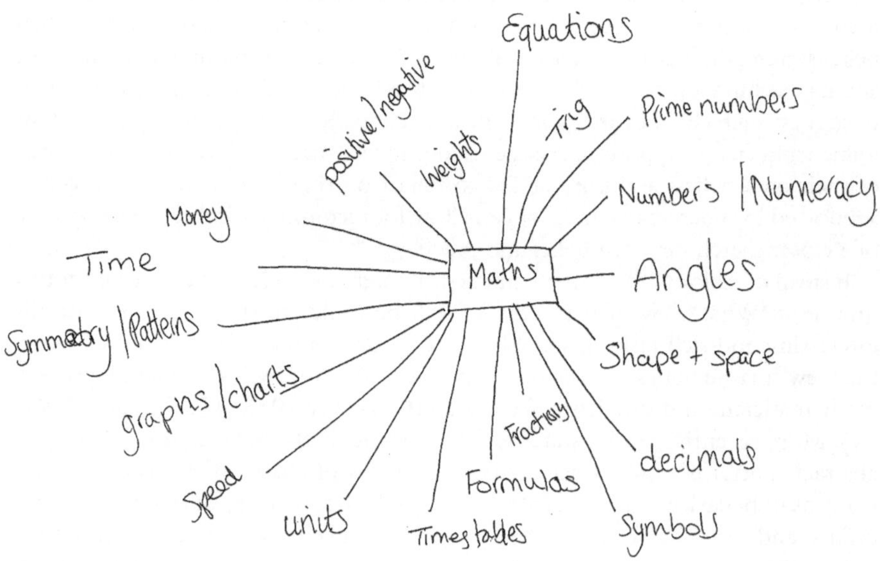

Figure 3.1a Concept map

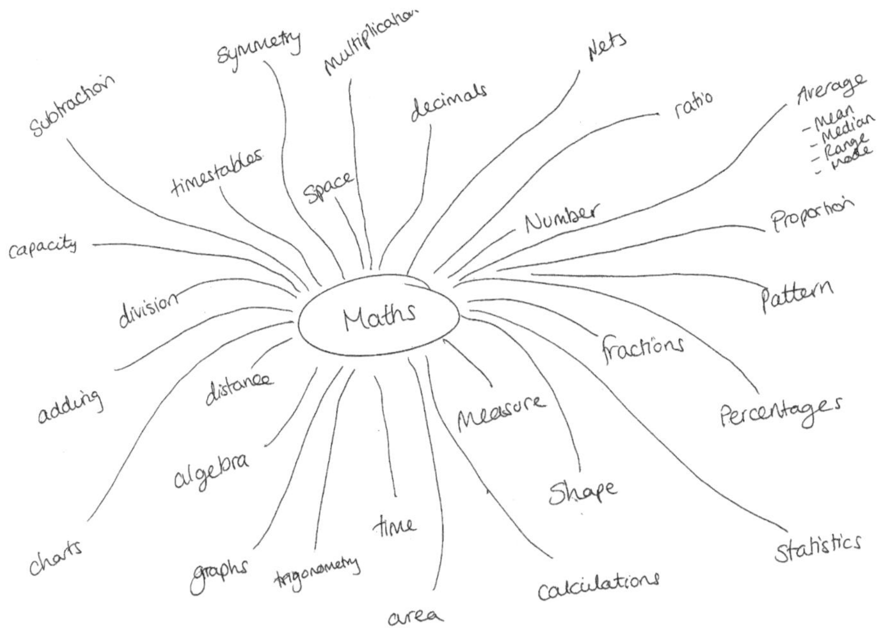

Figure 3.1b Concept map

tance of the link between numerical/algebraic and geometric content in school mathematics (Jones 2010), it is of concern that these concept maps still show pre-service teachers' limited views of the nature of mathematics.

Now, we turn to the view 'mathematics is a science of patterns', and see how focusing on 'patterns' can provide opportunities for productive learning processes, and develop an increasingly big picture of mathematics. Consider Pythagoras' theorem, which is often introduced as 'in a right-angled triangle, the area of square on the hypotenuse is equal to the sum of the areas of the squares on the other two sides'. Students simply see this as a 'rule' to remember. Students are then expected to undertake many exercises such as 'Find the lengths of the sides marked with a letter in the following right-angled triangles'. Alternatively, taking the 'mathematics is a science of patterns' point of view, the following learning process might be considered: (1) *First, prepare four congruent right-angled triangles* (note we can draw a right-angled triangle without knowing the length of a hypotenuse). (2) *Make new shapes by combining them freely.* Examples of this process are shown in Figure 3.2. (3) *Now, what is the length of the hypotenuse? Can you find it without using Pythagoras' theorem or measurement?* To do this, let us focus on squares. For example, by physically manipulating four triangles, the solutions shown in Figures 3.3 and 3.4 might be considered. (4) Finally, we can ask how we can generalise this pattern, and as expected, we can obtain the formula $a^2 + b^2 = c^2$, where a, b

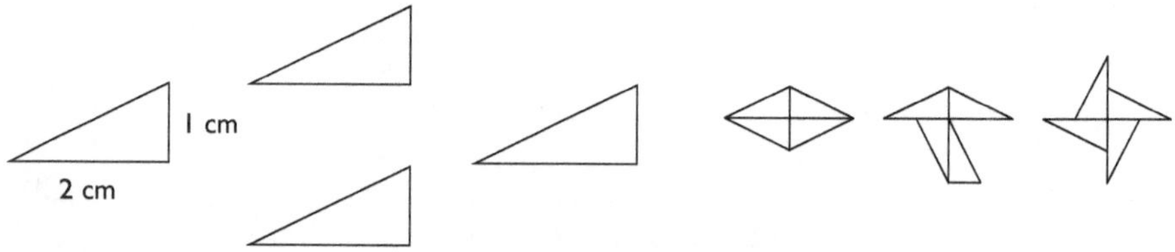

Figure 3.2 Making shapes by combining the triangles

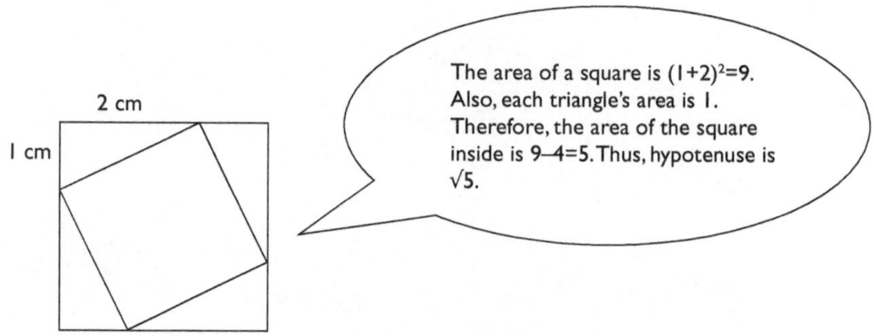

Figure 3.3 Solution 1

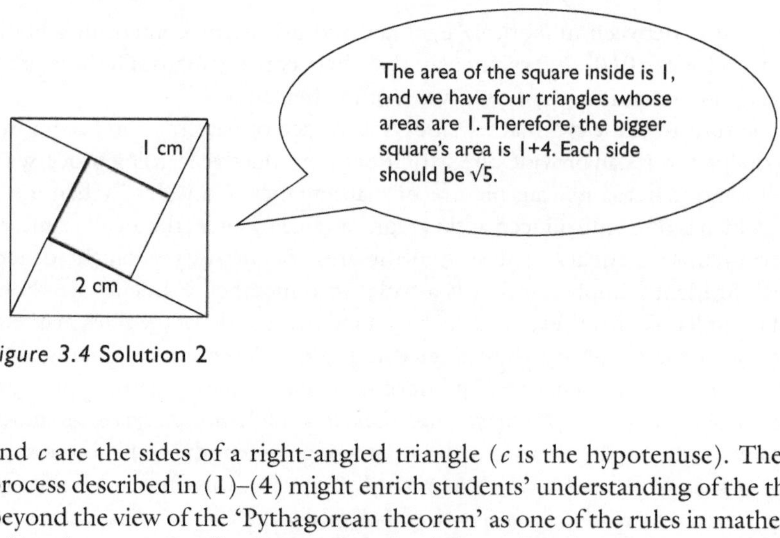

Figure 3.4 Solution 2

and *c* are the sides of a right-angled triangle (*c* is the hypotenuse). The above process described in (1)–(4) might enrich students' understanding of the theorem beyond the view of the 'Pythagorean theorem' as one of the rules in mathematics, taking it as a generalised pattern derived from the relationship between right-angled triangles, areas, squares (in both algebraic and geometric senses), triangles, and more importantly, the algebraic and geometric aspects of this theorem (see Yamamoto 2011 for further details of the activity).

Before undertaking problems such as these, it is useful to make a concept map, and then to revisit the concept map and modify it after trying the investigations. *Can you make new connections between elements of the concept map?* What new elements can be added? What mathematical processes are involved? Hopefully the original concept map gets bigger and bigger (both physically and conceptually).

What makes an effective teacher of mathematics?

Understanding the connections within mathematics and between mathematics and other domains is an important part of learning in mathematics. These connections can be between different aspects, topics, representations or structures in mathematics. These connections help learners to make sense of mathematics, to avoid misconceptions, make learning more interesting and help them to remember what they have learnt. Where teaching emphasises connections, new learning builds on existing understanding and problem solving becomes more effective as learners are able to draw on appropriate previous learning. Giving learners the opportunity to make connections for themselves and to see that new learning can be formed by connecting parts of previous learning allows learners to see that mathematics makes 'sense' and that it is not just 'plucked from the air'. One example is using prior learning about the area and circumference of a circle, along with that of cuboids and prisms, to find the volume and surface area of a cylinder. A connected approach to mathematics also allows learners to identify 'big ideas' in mathematics and provides cohesion to their mathematics learning. Problem solving also relies on learners looking for connections between a problem that is unfamiliar to them in some way and the mathematics they already know. Using connections to support problem solving also allows learners to make a choice from a range of different possibilities, select appropriate tools and materials to solve the problem and to develop their reasoning and justification skills.

Some pre-service teachers may never have thought explicitly about the connections within mathematics or may not be aware that these are not obvious to all. The design of schemes of work in schools tends to fragment the mathematics curriculum and separate it from curricula in other subjects. It is common for learners to spend perhaps two weeks working on a particular mathematics topic and then move on to a different topic that appears unrelated. Whilst there are good reasons for schemes of work to be designed in such a way, it can lead to learners finding it difficult to make connections between their mathematics learning and in recognising where the same technique is applied in a different context. Learners also often find it difficult to recognise instances where the same mathematics learning is used in other contexts, such as science or geography. Experienced teachers are skilled at helping learners to make connections between different learning experiences so mentors play a key role in supporting pre-service teachers in developing this skill.

In a simple, linear way this skill starts when pre-service teachers begin lesson planning. A key skill is that of deconstructing the topic to be taught so that it can be reconstructed in the classroom through a well-planned sequence of lessons where learning is effectively developed. This includes consideration of the pre-requisite skills required for aspects of the mathematical topic and a decision about the necessary starting points for the topic. Good mentoring can support pre-service teachers with developing a 'connectionist' view of mathematics by helping them to think about a topic in a range of ways including the other topics it connects to, the contexts it is used in and resources and activities that develop different approaches. Drawing a concept map for a topic in its early planning stages can be an effective way of identifying connections and relationships. Pre-service teachers should also be encouraged to include opportunities for learners to explore the connections within and between mathematics topics and to make these explicit.

Research by Askew *et al.* (1997) indicated that teachers with what they describe as a *connectionist* approach to mathematics and mathematics teaching and learning were more effective than those with a *transmission* or *discovery* approach. The report contains a table summarising the characteristics of teachers holding each of the three approaches. This table provides good discussion opportunities for a pre-service teacher and their mentor and might provide challenges to the pre-service teachers' notions of 'good mathematics teaching'.

Big ideas in mathematics

What do you think are the 'big ideas' in mathematics? Are they the same for primary and secondary school mathematics? Randall (2005) suggests that big ideas are those that are central to the learning of mathematics and those that link "numerous mathematics understandings into a coherent whole" (p. 10). It might well be an instructive undertaking for mentor and pre-service teacher to each write their own list of those things they can identify as the big, underpinning, principles of school mathematics. Comparing lists then provides an opportunity to discuss some of the long-term goals teachers have for mathematics learners.

A quick search on the Internet will reveal a number of suggested lists of these 'big ideas'. Randall (2005), for instance, suggests 21 big ideas in mathematics in his list but Faux (1998) has only six, along with three big ideas in mathematics education. ACME, the Advisory Committee on Mathematics Education for England, suggests that the big ideas in primary mathematics should include place value and the number system, conservation of number and measure, equivalent relations and dimensionality (ACME 2008). They suggest that these big ideas are often overlooked "in favour of specific lesson objectives that are measureable and easily assessable". For mentors and pre-service teachers, then, big ideas provide an opportunity to discuss planning more cohesively and in a more long-term manner for students' learning beyond a day-to-day approach. A view of teaching and learning mathematics that encompasses some big ideas also complements a view of mathematics as a connected body of knowledge.

The Ontario Ministry of Education (2008) suggests that using big ideas can help teachers in:

- making instructional decisions;
- identifying prior learning;
- looking at students' thinking and understanding in relation to the mathematical concepts addressed in the curriculum;
- collecting observations and making anecdotal records;
- providing feedback to students;
- determining next steps;
- communicating concepts and providing feedback on students' achievement to parents.

(p. 16)

Such a list provides important practical justification for a 'big ideas' approach to teaching mathematics and provides opportunities for mentors and pre-service teachers to discuss learners' progress more holistically.

Developing mathematically rich tasks

The design of mathematically rich tasks is one of the most important challenges for mathematics educators, and again various views and beliefs influence task design. For example, if a person holds an *instrumentalist* view ('mathematics as a set of rules'), then for this person a mathematically rich task might be a set of exercises so that learners can see and test rules in mathematics. On the other hand, for a person with a *problem solving* view, a set of exercises using routine algorithms is unlikely to be enough. But how can we design mathematically rich tasks, and how can mentors working with pre-service teachers support them to develop their skills designing these tasks?

We have already discussed the view that 'mathematics as a science of patterns' can be a useful starting point, but is it enough just to pay attention to mathematical patterns? For example, number sequences have various patterns, but does the following task contain 'mathematically rich' learning opportunities?

Find the nth term of the sequence U_n: 1, 3, 5, 7, 9, . . .

While the task above does provide us with an opportunity to generalise a sequence which is certainly important in the learning and teaching of mathematics, it is still a simple exercise about arithmetic sequences. It also shows us that the use of patterns might not be enough to design mathematical tasks in which learners can explore and appreciate various relationships between mathematical concepts. But again, how can we conceptualise mathematically rich tasks? To address this issue, we introduce an idea from Wittmann's (1995) Substantial Learning Environment (SLE). He states that an SLE has the following key features:

1. It represents key objectives, content and principles of teaching;
2. It is related to significant mathematical content, processes and procedures beyond this level, and is a rich source for mathematical activities;

3 It is flexible and can be adapted to the conditions of particular classrooms;
4 It incorporates mathematical, psychological and pedagogical aspects of teaching mathematics in a holistic way, and offers a wide potential for empirical research.

(Wittmann 1995: 365–6)

The underpinning of what makes Wittmann's SLEs a rich source of mathematical content is the view of 'mathematics as a science of patterns'. Let us take some examples, and show how we can utilise the above four points as a 'checklist', i.e. if a task contains the above four features then this task can be seen as 'mathematically rich'.

Let us take an example, called Bamboo numbers, which is based on the patterns of the Fibonacci sequence, but made by a set of finite recursive relations, e.g. 'Based on the rule 1-3-4-7, complete 5-1-?-?, 2-4-?-?, 6-0-?-?, 4-2-?-?, 1-?-?-11 and so on.' A possible representation of this problem as an activity for learners is shown in Figure 3.5 and the completed activity is shown in Figure 3.6.

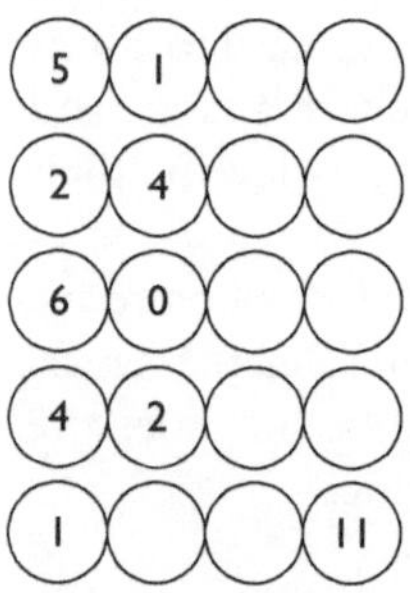

Figure 3.5 **Bamboo numbers activity**

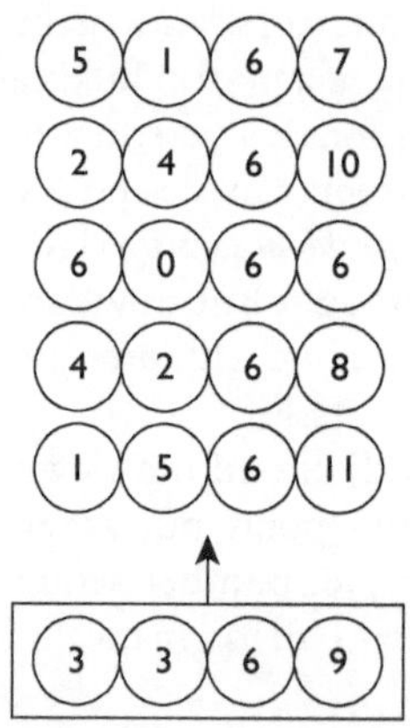

Figure 3.6 **Completed activity**

The 'Bamboo numbers' activity satisfies the four key features of SLEs. Through working on 'Bamboo numbers', children notice that these numbers form some patterns, e.g. natural numbers are used, the last numbers are consecutive (6, 7, 8, 10 and 11) except 9, the third number is the same, the pattern is completed if 3-3-6-9 is added etc., which are all derived from the properties of the Fibonacci sequence (key features (1) and (2)). We can also extend these patterns by considering, for example, the following questions: If we use five numbers, then what happens to the last numbers? Can we use decimals or fractions, but still enable the last number to be consecutive? If the first number is less than zero, what would happen? Also, without changing mathematical structures of this task, it is possible to make it easier for lower attaining learners (key feature (3)). Mason, Graham, Pimm and Gowar (1985) propose four roots of algebra, 'Generalized Arithmetic', 'Expressing Generality', 'Possibilities and Constraints', and Rearranging and Manipulating'. 'Expressing Generality' is considered as "the heart of mathematical thinking" (ibid., p. 8), and consists of the following four stages: 'Seeing', 'Saying', 'Recording patterns', and 'Testing formulations'. The Bamboo numbers activity offers such opportunities for algebraic thinking for primary school children early on in their schooling. Finally, the mathematical thinking that can be observed when children are undertaking the Bamboo number task is worth studying in mathematics education research (key feature (4); see, for example, Fujita and Yamamoto 2011).

We acknowledge the ideas described above are not new; indeed, this type of problem solving task has been central in recent mathematics education. But from the 'mathematics as a science of patterns' point of view, learners not only investigate mathematical patterns which are intended in given tasks, but also create their own patterns beyond what teachers expect them to study. Wittmann (2006) himself extended the science of patterns to that of 'dynamic patterns':

> What matters is not the science of ready-made and static patterns but the vital science of dynamic patterns which can be developed globally in the curriculum as well as explored, continued, re-shaped, and invented locally by learners themselves.
>
> (2006: 2)

In the Bamboo numbers activity, let us focus on the pattern 2-4-6-10, and regard the first three numbers as a '3-digit number', 246, where 2 + 4 = 6 and 4 + 6 = 10. The learner might notice that in adding 191 something similar emerges, i.e. 246 + 191 = 437, as 4 + 3 = 7 and 3 + 7 = 10. Repeat this, 437 + 191 = 628 and so on, and one can create patterns quite easily. Moreover, in the case of the patterns which consist of four numbers, the last number does not have to be 10, and many other patterns are easily created by adding 191 (the above 'pattern' was discovered by an 11–year-old boy) as shown in Figure 3.7.

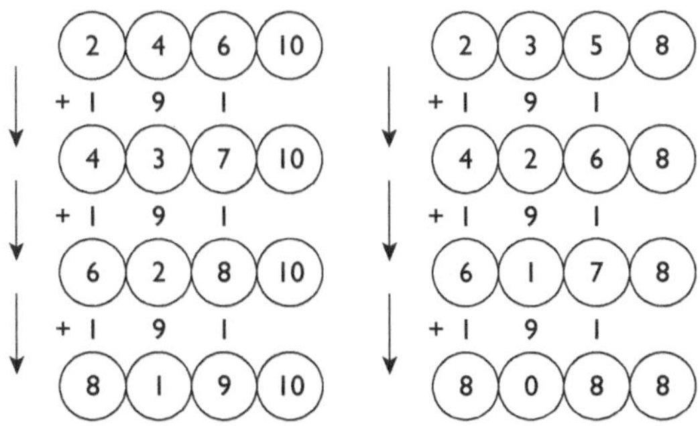

Figure 3.7 Adding 191 pattern

The view of 'mathematics as the science of patterns' and the criteria for SLEs can provide a useful starting point and checklist for mathematically rich tasks. (In Germany, these ideas have been implemented already in primary schools, e.g. Mueller & Wittmann 2004.) Mentors can use the four features to check whether tasks designed by pre-service teachers are mathematically rich or not. Next we extend this approach and consider the development of learning environments for the use of daily lessons in schools.

Developing learning units

Once you have a good task, then it is necessary to develop the task further for use in the classroom, i.e. by designing learning units or sequences of lessons from individual tasks. Our first context is 'Race to 15', which is a variation of 'Race to 20' described by Wittmann (2001) (see Figure 3.8). This is a game for two, and the rules are as follows:

- The first player starts by putting 1 or 2 counters on the first circle, or the first two circles.
- The second player follows by putting 1 or 2 counters down.
- The players take turns until one of them arrives at the target number 15 (winner)

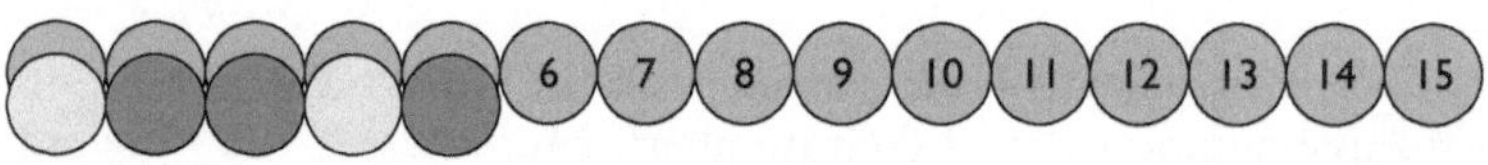

Figure 3.8 Race to 15

After playing this game several times, you soon notice that you need to get 12, 9, 6, and 3, i.e. the second player will always win. This task satisfies the four key features of the SLEs. First, by undertaking this task learners can see relationships between counting, adding, subtracting, multiples, and so on which are all important content in current school mathematics. Also, the process for identifying the winning strategies involves trial and error, justifying testing, and so on, which are related to important mathematical thinking (key features (1) and (2)). This task is very flexible as you can change the target number and the numbers of counters, and can be adapted for a range of age groups from primary to secondary school (key feature (3)). Wittmann (2001) also reported that there have been several research studies into children's learning processes when exploring the patterns and structures in this task (key feature (4)).

In general, teaching units with the SLEs can start from several simple cases and relationships, and then extend to more general cases and more general patterns, i.e. the lesson sequences can simply follow how mathematical patterns are revealed by learners. In the case of 'Race to 15', it can be used both in primary and secondary schools. For example, primary school children can work with particular target numbers and counters. They can start from 'Race to 15' or 'Race to 10' for lower attaining learners. Once they grasp the rules of the game, you can then discuss the winning strategies of this case, and then suggest that learners change the target number. You could stop at the case of the number of counters as 2, but of course the numbers of counters can be investigated as well! Students in secondary schools can investigate this game further, e.g. investigating a general winning strategy when the target number and the number of counters are 'm' and 'n', respectively.

Our second context is arithmetic sequences. We have seen that giving number sequences individually might not be enough to be a mathematically rich task. Students might also then see the study of number sequences as learning many formulae about sequences and series, and see little relationship between them (the *instrumentalist* view). To make the study of number sequences a rich source of mathematical learning, might studying sequences and series together be an interesting idea? For example, let us take the following example:

> Express natural numbers as a sum of consecutive whole numbers, such as $6 = 1 + 2 + 3$, $18 = 3 + 4 + 5 + 6$, etc.

Again we believe this simple problem satisfies the four features of the SLEs. On undertaking this task, you will quickly see many relationships and patterns between numbers which learners in both primary or lower secondary schools can explore. For example, even if students simply investigate several numbers there are still many patterns they can investigate. For example:

- some numbers cannot be expressed as the sums of consecutive whole numbers, e.g. 2, 4, 8, 16, etc. (2^n)
- multiples of 3 (>3) can be expressed as the sum of 3 numbers, e.g. $9 = 2 + 3 + 4$, $15 = 4 + 5 + 6$, etc. Multiples of 5 (>10) can be written as 5 numbers, e.g. $15 = 1 + 2 + 3 + 4 + 5$, $20 = 2 + 3 + 4 + 5 + 6$, etc.
- triangular numbers 1, 3, 6, 10 etc. can be expressed as $1 + 2 + 3 + 4 + \ldots$

In their mathematics lessons, learners can first investigate numbers from, say, 1–32, then focus on their findings, and then take a few examples such as considering the multiples of 3. Upper secondary school students can pursue these patterns further. They can use algebra to explain the cases of the multiples of 3, 5, etc. By using the formulae of arithmetic sequences and series, they can prove why the powers of 2 cannot be expressed as a sum of consecutive numbers and investigate how to find answers for the given numbers such as 496.

The role of the teacher

While the idea of an SLE can provide learners with rich mathematical learning opportunities, the role of the teacher in managing teaching should not be ignored. Mason (1998) notes that "the role of a teacher is to create conditions in which students experience a corresponding shift in the structure of their attention, in which they become aware of acts and facts of which they were previously unaware" (1998: 244). We have also discussed that a *connectionist* approach would be more effective than other approaches. Considering the effectiveness of this approach, and to make learners aware of mathematical patterns in the SLEs, it is also important to facilitate learners' mathematical thinking so that they are able to make links between various mathematical patterns and structures. Once the tasks and teaching units are designed, it is useful for both mentors and pre-service teachers to discuss and exchange ideas for when and how to give key instructional interventions to promote learners' mathematical understanding.

For example, suppose children are undertaking the following Bamboo number task (Figure 3.9). To complete 10-?-?-?-53, it is useful to pay attention not only to horizontal but also vertical relationships between numbers (Figure 3.9). However, is just saying 'Please pay attention to how each number increases!' enough to solve this problem? Our experience with implementing the SLEs implies the following: Children cannot solely grasp all the mathematical patterns and structures of Bamboo numbers by themselves; even though the teacher gives instructions to pay attention to investigate vertical and horizontal relationships, not all children (8 years old) could solve the problem shown in Figure 3.9 (Fujita and Yamamoto 2011).

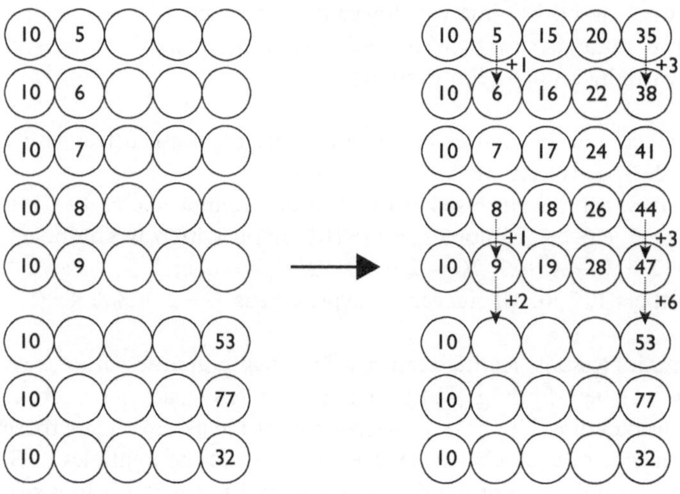

Figure 3.9 Bamboo number problem

Now, how do you approach facilitating children's thinking so that they are able to solve this problem? How can we help children see the relationship from a holistic point of view? Again, it is useful for mentors and pre-service teachers together to refer back to *connectionist* views and big ideas of mathematics mentioned in the previous sections and to list useful tactics to identify the roles of the teacher in mathematics lessons.

Final remarks

Beliefs about the nature of mathematics, how mathematics is learnt and the role of the teachers are significant influences on how pre-service teachers approach learning in mathematics. Identifying and challenging differing perspectives is an important part of understanding how learning is viewed by pre-service teachers and their mentors. We are of the view that belief in mathematics as 'the science of patterns' and of mathematics as a connected body of knowledge containing a small number of 'big ideas' allows for the development of powerful learning opportunities for students of all ages. When mentors and pre-service teachers plan rich tasks that meet Wittman's (1995) criteria for a Substantial Learning Environment, and mentors are able to help pre-service teachers to manage learning through the use of such tasks in the classroom and to identify the key role they play in this process, learners have opportunities to engage in their mathematics meaningfully and purposefully. This, surely, is a sound goal for mathematics learning in schools.

References

Advisory Committee for Mathematics Education (2008) *Mathematics in Primary Years: A discussion paper for the Rose Review of the Primary Curriculum*. Online. Available HTTP: <http://www.acme-uk.org/media/1769/rose%20review%20maths%20paper%20final%20sept%2008.pdf> (accessed 18 October 2012).

Askew, M., Brown, M., Rhodes, V., Johnson, D. and Wiliam, D. (1997) *Effective Teachers of Numeracy: Final Report – report of a study carried out for the Teacher Training Agency 1995–96 by the School of Education, King's College London*, London: King's College.

Devlin, K. (1998) *The Language of Mathematics*, New York: W.H. Freeman and Company.

Ernest, P. (1989) 'The impact of beliefs on the teaching of mathematics', in P. Ernest (ed.), *Mathematics Teaching: The State of the Art*, New York: Falmer, pp. 249–53.

Faux, G. (1998) 'What are the big ideas in mathematics?' *Mathematics Teaching*, 163, 12–8.

Feynman, R.P. (1968) 'What is science?' *The Physics Teacher*, 7(6), 313–20.

Fujita, T. & Yamamoto, S. (2011) 'The development of children's understanding of mathematical patterns through mathematical activities', *Research in Mathematics Education*, 13(3), 249–67.

Hardy, H.G. (1940) *A Mathematician's Apology*, with a foreword by C.P. Snow, Cambridge: Cambridge University Press. Repr. Canto Edition, 1992.

Jacquette, D. (ed.) (2002) *Philosophy of Mathematics: An Anthology*, Oxford: Blackwell.

Jones, K. (2010) 'Linking geometry and algebra in the school mathematics curriculum', in Z. Usiskin, K. Anderson, & N. Zotto (eds), *Future Curricular Trends in School Algebra and Geometry*, Charlotte, NC: Infoage, pp. 203–15.

Mason, J. (1998) 'Enabling teachers to be real teachers: Necessary levels of awareness and structure of attention', *Journal of Mathematics Teacher Education*, 1(3), 243–67.

Mason, J., Graham, A., Pimm, D., & Gowar, N. (1985) *Routes to, Roots of Algebra*, Milton Keynes: The Open University.

Mueller, G. & Wittmann, E.Ch. (2004) *Das Zahlenbuch* Mathematics for the Primary Grades Vol. 1–4, Liepzig: Klett.

Ontario Ministry of Education (2008) *Patterning and Algebra Grades 4-6: A Guide to Effective Instruction in Mathematics Kindergarten to Grade 6*, Ontario: Queen's Printer for Ontario.

Randall I.C. (2005) 'Big ideas and understandings as the foundation for elementary and middle school mathematics', *Journal of Mathematics Education Leadership*, 7(3), 9–24.

Sawyer, W.W. (1995) *Prelude to Mathematics*, London: Penguin.

Wittmann, E.Ch. (1995) 'Mathematics education as a "Design Science"', *Educational Studies in Mathematics*, 29(4), 355–74.

Wittmann, E.Ch. (2001) 'Developing mathematics education in a systemic process'. *Educational Studies in Mathematics*, 48(1), 1–20.

Wittmann, E.Ch. (2006) 'Mathematics as the science of patterns – a guideline for developing mathematics education from early childhood to adulthood', *Annales de Didactiques et de Sciences Cognitives*, 11, 149–74.

Yamamoto, S. (2011) *A Pattern in a Right Angled Triangle*. Online. Available HTTP: <http://www.educ.kumamoto-u.ac.jp/~shinya/Reasarch%20Activities/Research%20Activities%20Abroad%20/2011%20Plymouth/Worsksheets/A%20pattern%20HIwaki.pdf> (accessed 21 November 2012).

Part 2

Equity and diversity issues

Chapter 4

Learning to teach mathematics in a multicultural society

Manjinder Jagdev and Avril Mallett

Mathematics is culturally situated, with its own history that has developed in rich and diverse ways in different cultures over thousands of years, and there are benefits to young people in experiencing mathematics in these ways. The learning and teaching of mathematics can provide opportunities for practice-based mentors and pre-service teachers to show that they value diversity and challenge racism in the classroom. This chapter offers school-based mentors further consideration of these issues, models of support for developing such approaches with beginning teachers and what might be meaningful learning experiences for students using historical and cultural aspects of mathematics.

Introduction

The learning of mathematics has underpinned the foundations of education in every society, at all levels, for centuries. Whilst few doubt its importance, over time the content of school mathematics has changed and developed, both in response to society's perceived needs, and as understandings about the nature of mathematics and its learning have developed. However, valuing historical and cultural approaches to mathematics teaching is dependent on one's own understanding of the nature of the subject. Many beginning teachers have little personal knowledge about the historical and cultural roots of mathematics and this is also an area of unfamiliarity for many practising teachers acting as practice-based mentors to these beginning teachers. Understanding how mathematics can contribute to a multicultural and anti-racist education for all young people can therefore become a challenge for all those involved. Banks (1992: 90) supports this view saying, "multicultural education . . . changes must be made . . . including mathematics and science . . . Much more important work needs to be done to provide teachers with examples and specifics they need . . . we can help all teachers, including mathematics and science teachers."

Content, nature and learning of mathematics

Mathematics was regarded as the purest of all disciplines in Hindu society, contained in the Vedas (holy books); Brahmins, members of the highest caste in medieval times, saw mathematics as holding the key to the universal mysteries of the stars and planets and timeless truths of the gods. They spent their lives studying and advancing mathematics. Vedic mathematics covered all aspects including calculus and is the oldest written record of mathematics.

Development of mathematics in the Greek and Roman world related to the kind of societies they were. European dominance came from Pythagoras. Pythagoras' theorem and Euclidean geometry solved contemporary problems in building Greece, Rome and Egypt. Research has revealed references by the Greeks themselves to visits by their great mathematicians such as Thales and Pythagoras to Egypt, Babylon and India to study mathematics. Furthermore, the *Cockcroft Report* (1982) affirmed that mathematics should be used to impart a variety of information about everyday life, and in the past, curriculum changes have prepared students for various industries and economic demands.

The truly international membership of the International Study Group on the Relations Between the History and Pedagogy of Mathematics suggests that there is recognition of the importance of multicultural and historical approaches to teaching mathematics across the world. For example, such approaches are embedded in the mathematics national curriculum (DfE 2007) for England and Wales. The section about the curriculum for 11–14-year-olds states the significance of "Recognising the rich historical and cultural roots of mathematics" and that "Mathematics has a rich and fascinating history and has been developed across the world to solve problems and for its own sake. Pupils should learn about problems from the past that led to particular areas of mathematics" (2007: 141) and "The subject transcends cultural boundaries and its importance is universally recognised" (2007: 145).

Culturally situated mathematics and its benefits

In the third edition of his influential book *The Crest of the Peacock* (2011), George Gheverghese Joseph provides alternative perspectives on the history of mathematics. With a readership of mainly teachers, this important resource outlines the non-European roots of mathematics. From the 'Mathematics from Bones, Strings, and Standing Stones' of Africa and the 'Mayan Numeration' to the dawn of modern mathematics, this multicultural tour shows us the deep influence that the Egyptians and Babylonians had on the Greeks, the successes of Chinese and Indian mathematics, as well as the Arabs' major contributions. This volume explores how mathematical ideas travelled from East to West, illuminating the global history of mathematics.

Activities can be planned from the book's wide scope, including this example based on Counting Systems and Numeration: The Pacific Dimension, for primary school students.

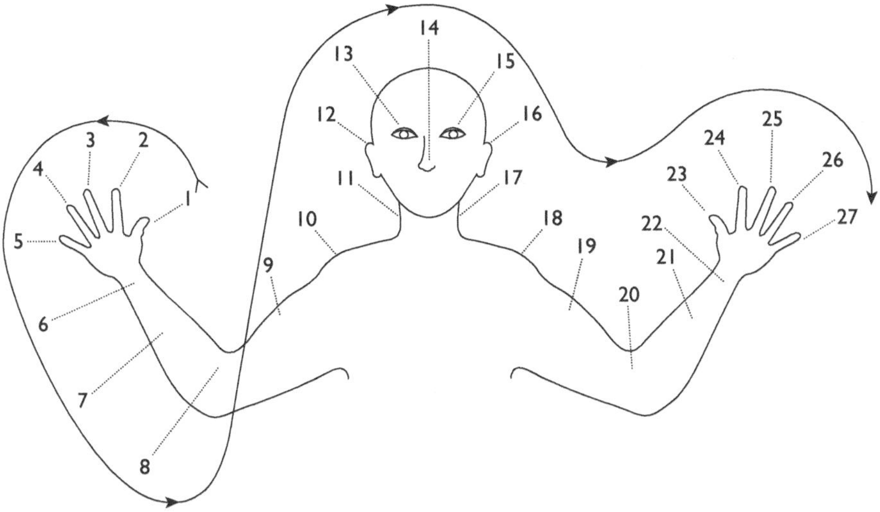

Figure 4.1 **Body part tally system adapted from Saxe & Esmode (2005)**

Primary school children can be taught the 'body part' tally system where numbers are represented on the body and counting on body parts has a conventionally defined order. Saxe describes the counting system of the Oksapmin, a group found in the West Sepik Province of Papua New Guinea, according to Figure 4.1 (Saxe and Esmonde 2005). Primary school children can repeat the counting process again to reach 54 on the second count with a specific number referring to the relevant part on the body. They can apply this to counting in a number of kinaesthetic activities, for example, a set of model farm animals, measuring the length in centimetres of their classroom desks, or locating their classroom in relation to other classrooms along the corridor.

Primary school beginning teachers can plan activities for their students involving foreign currency, looking at the values of coins from different countries. Simple probability games using these coins can be played with students making a colourful display of coin rubbings. A greater understanding of other cultures can be gained from international links between schools and exchange trips for both teachers and students.

Secondary school mentors and beginning teachers can use Islamic Art from ancient Baghdad to teach geometry and symmetry. On his visit to Pakistan and the Naulaka Pavilion in Lahore, Michael Palin (2005: 57) says: "The view from this cool pavilion out towards the Badshahi Mosque is a reminder of what makes Mughal architecture so fine. It's all about balance and symmetry. Towers, domes, minarets, columns and cupolas, some in red stone, some in white marble, are all gracefully harmonized." Windows, grilles and screens, and Moorish doorways built in Islamic times in Spain are rich material for beautiful circles, hexagons and rectangles, triangles, five-pointed stars and more. Designs, ideas and diagrams, such as the one in Figure 4.2, for students to copy and practise their compass skills are to be found in the magical book *Islamic Patterns* by Keith Critchlow (1976).

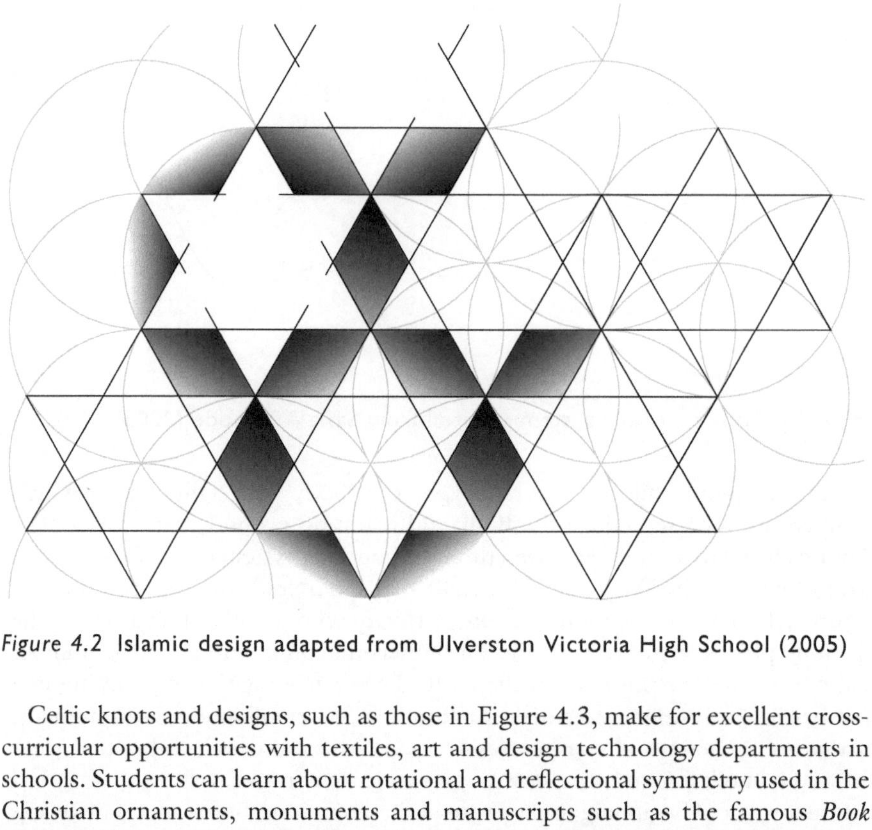

Figure 4.2 **Islamic design adapted from Ulverston Victoria High School (2005)**

Celtic knots and designs, such as those in Figure 4.3, make for excellent cross-curricular opportunities with textiles, art and design technology departments in schools. Students can learn about rotational and reflectional symmetry used in the Christian ornaments, monuments and manuscripts such as the famous *Book of Kells*. The wonderful motifs in the book *Celtic Iron-On Transfer Patterns* by Courtney Davis (1989) make for active mathematics in the secondary classroom.

Student feedback from one of the authors' previous research, 'Educational perspectives from Black and Asian Pupils in an inner-city multicultural comprehensive' (Jagdev 1996) included '*more black history*'; students wanted to know '*where maths was discovered*'. With stimulating resources, students can develop an appreciation of mathematics from many cultures. In my own classroom, students were amazed to learn that the Hindus invented zero. Thus, historical and multicultural activities also give opportunities to challenge stereotypes of mathematics being Euro-centric.

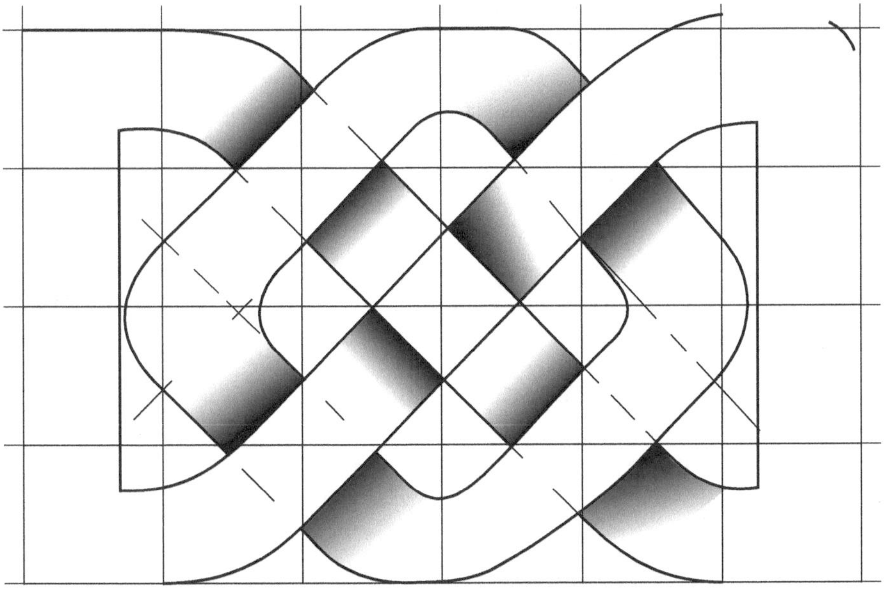

Figure 4.3 Celtic knotwork adapted from Ulverston Victoria High School (2005)

Valuing diversity, developing mathematical understanding and challenging racism

Classroom activities based on historical and cross-cultural roots of mathematics also relate to citizenship issues. The racist murder of Stephen Lawrence took place on 22 April 1993 in southeast London. Stephen, son of Doreen and Neville Lawrence, and older brother of Stuart and Georgina, was an 18-year-old gifted athlete who trained and competed with his club. After eating dinner with Stephen's uncle, Martin, and Martin's wife Millie, Stephen and his friend Duwayne walked to a bus stop, after 10.35pm, to take them in the direction of their respective homes. Here, Stephen was stabbed twice by a group of white youths. As Duwayne managed to escape the attack, Stephen, despite his injuries, was able to run up the hill towards Duwayne, but fell and died due to excessive bleeding.

The powerful text *The Life of Stephen Lawrence* by Verna Wilkins (2004) provides further contextual detail for mentors and beginning teachers. This book, published as primary school children's non-fiction, sensitively treats the tragedy and celebrates Stephen's development as a young man with a passion for music, art and mathematics, and loved by family and friends.

Nineteen years later, two criminals responsible for Stephen's murder were convicted. Doreen Lawrence can be heard discussing the 'guilty verdict' of two of the murderers of her son on BBC Radio 4 Woman's Hour (Lawrence 2012). The Macpherson Report (1999), an enquiry highlighting institutional racism in

UK police, stemmed from Stephen's murder and brought changes to police practices in the UK. The report makes important recommendations for the role of education. These include the provision of citizenship education aimed at valuing cultural diversity and preventing racism, in order better to reflect the needs of a diverse society. The Stephen Lawrence enquiry also recommends that relevant agencies should specifically consider implementing community and local initiatives aimed at promoting cultural diversity and addressing racism and the need for focused, consistent support for such initiatives.

Beginning teachers and mentors can use the following race statements by asking students to consider citizenship issues and challenge racism when identifying the different ethnicities of those referred to below:

> I am twice as likely to die by the age of one;
> I am half as likely to grow up in poverty;
> I am three times as likely to be excluded from school;
> I am three times as likely to go to a leading university;
> I am four times as likely to be murdered;
> I am one fifth as likely to end up in jail;
> I am three times as likely to be poor in old age;
> I am likely to be paid 20% more for the same work.
>
> (adapted from Vallely, 2012)

Parekh (2000) recommends that citizenship education should include human rights principles, skills on advocacy and campaigning, equality legislation and opposition of racist beliefs and behaviour; it should teach about cultural diversity. In England, the Diversity and Citizenship Curriculum Review (DfES 2007) focuses on valuing cultural diversity, race equality, multicultural school policies, developing a more tolerant society, and exploring identity and national identity. It highlights the importance of emphasising the contributions of mathematics to education for diversity and provides two examples of contextualising mathematics in schools. Mentors and beginning teachers can discuss these recommendations and consider the issues of diversity and communities, texts in mathematics, how we live together and deal with difference, the role of mathematics in citizenship education, and individual support for students.

Beginning teachers can jointly plan lessons with the book *Human Rights in the National Curriculum* by Peter Wright (2004) to teach mathematics and citizenship simultaneously. This book addresses human rights and challenges of citizenship education within the context of the secondary mathematics classroom with thirty multicultural activities that have been developed by teachers. These have been fully trialled in a range of schools. Each activity is accompanied by teachers' notes and a resource sheet for photocopying, including number and calculation – Yoruba numbers; games and puzzles – magic dominoes; and handling data – fair trade chocolate. Later in this chapter, an experienced mentor describes her use of Amish quilt design to teach shape and space.

One example suitable for the primary classroom comes from the resource *Mathematics from Many Cultures* by Claudia Zaslavsky (1995). Alquerque is the Spanish game which originally came from North Africa. It is still a very popular game in Spain and the modern game of draughts is based on alquerque. This is a game for two players, where each has twelve counters, as shown in Figure 4.4. Different versions of the game are played by children, including Awithlaknannai, played by the Zuni people of New Mexico, and Peralikatuma, played in Sri Lanka. The latter has the same rules but the board has four additional triangles and each player has 23 counters. Students can be asked to make comparisons between the boards.

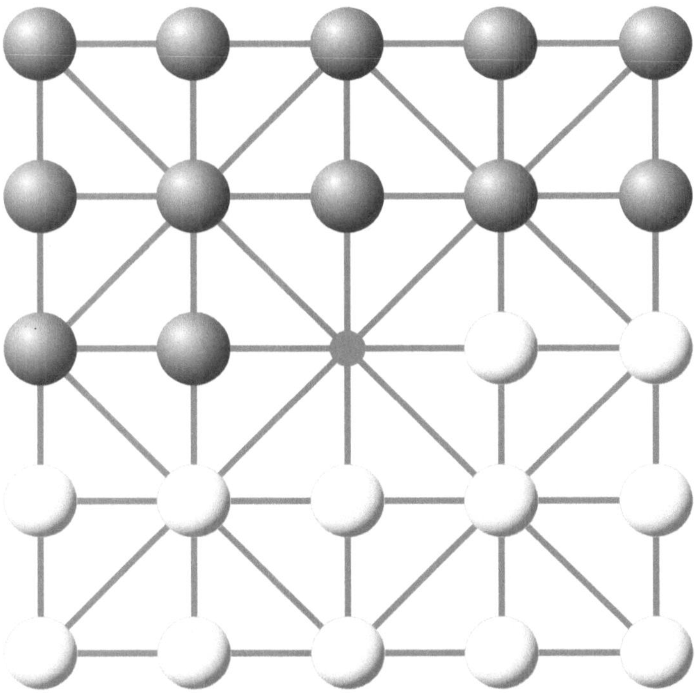

Figure 4.4 Alquerque board and counters

Another example from this colourful material encourages students to consider the cubit, used by the Ancient Egyptians to measure length. Students measure their cubit – the distance from the elbow to the tip of the middle finger – and their handspan – the distance across an outstretched hand from the thumb to the little finger. They can then be shown how to use their handspan when measuring cloth to sew or knit in the same way that my mother used her handspan, incredible spatial awareness and symmetry skills to design and make Indian shalwar-kameez outfits. Once students are introduced to the metric system, developed by French scientists during the 1790s, they can make a list of a variety of objects that are about the same length as a kilometre, hectometre, decametre, decimetre, centimetre and millimetre.

At the end of the lesson using such material, as part of a plenary, mentors and beginning teachers can pose the following questions to learners: 'What did you find easy / hard / uninteresting / enjoyable?' Another plenary could be for learners to answer: 'One thing I already knew today . . . / two questions about what I have learnt or two things I don't understand . . . / three things that I learnt today . . ..'

Tariq Modood (2007: 45) states that "The most important cultural racism today, at least in western Europe, is anti-Muslim racism, sometimes called Islamophobia. A multicultural approach, recognizing the plurality of racisms and the distinctive needs and vulnerabilities of different groups, is therefore what is needed to tackle racial and religious discrimination." School mentors can utilise the services of professionals from local community organisations to visit their schools and discuss, with both teachers and learners, how to challenge racism.

Pre-service teachers, mentors and multicultural mathematics

Following a session about the historical and cultural roots of mathematics with pre-service teachers, comments included:

> I am interested in finding out more about different teaching styles (of mathematics) from different countries.
>
> Didn't realise the extent to which it could be used in lessons.
>
> I love Islamic patterns and geometry of Islamic buildings. I visited the Taj Mahal once and was overawed by it. I would love to bring that experience into the classroom, and your lesson gave me some insight into how I might do that.
>
> There are some lovely historical stories linked to maths, such as Fermat's Last Theorem – its creation and then (recently) its proof; the feud between Isaac Newton and Liebnitz [sic]; Einstein and his work . . . this could be a great x-curricular topic! It's a pity there is so little time to explore these ideas on our course.
>
> Likewise, practising multiplication or similar with a cultural twist is good. The more versatile it is, and the more suggestions for topics it could be weaved into, the better. Overall I am keen to incorporate historical and cultural roots of maths into my teaching but I am not sure how to do it or when I'd have time to research activities and info so a session like yours is v helpful to get the ball rolling.

Models of support and good learning experiences

In a session on multicultural mathematics with pre-service teachers, I highlight three activities: multiplication tables in different languages; Vedic mathematics; and global statistics from the publication *New Internationalist.* The first two have been taken from the excellent resource book *Multiple Factors: classroom mathematics for equality and justice* by Sharan-Jeet Shan and Peter Bailey (1994).

Exploring multiplication tables in different languages allows for a celebration of different cultures in the classroom. Keywords include 'product' for students to define in class; this is an excellent activity for learners to practise mental arithmetic skills. Simply give out the multiplication squares, with symbols 0 to 9 as indicated for a variety of languages, to students for completion. For lower-attaining learners, encourage them to write the Roman numerals 0 to 9 alongside the symbols as a first activity. As an extension, students can create sums, facts and questions using the symbols of the languages explored such as: 'My age is . . .'; 'I live at number . . .'; ' = ٨٥ + ٢٩ '; 'What is your shoe size?'; ' = ٣ multiplied by ٤ '. Students' posters can then be put on display for other students to answer the questions.

Pre-service teachers commented:

> The multiplication grids in other languages. Using them would be an effective way of celebrating diversity whilst still allowing the children the chance to use mathematical and problem-solving skills. Ideas on how to use, e.g. Islamic patterns in maths lessons could be useful.
>
> I liked the multiplication tables in different languages – the benefits for practising multiplication tables for weaker pupils, experiencing difficulties that EAL pupils may have.
>
> Multiplication tables could be used in a wide range of lessons – helped me think of ways to improve my practice.
>
> I really enjoyed creating a number table using numbers from a different culture. It gave an insight into what it must be like to be an EAL learner.
>
> . . . Intend to use it for multiplication revision and use the similarities to talk about maths being created or existing.
>
> . . . Fun to do, wanted to finish when time ran out.

One pre-service teacher completed a multiplication table in Braille! This activity generates discussion of languages where script is written from right to left and whether this affects the writing of numbers with more than two digits. As an extension, students can be encouraged to create their own symbols for number systems. A student commented on her piece of work: 'I have learnt that some Punjabi numbers look similar to the English ones!' Thus classroom discussions can take place about how the Roman numerals used today have evolved from medieval Islam's influence in the spread of Indian numerals, through history.

Whilst engaged in this activity in the classroom, a student asked me about my language and explained her first language as Bengali (a multiplication grid for which can also be found in Shan and Bailey's book). When I told her that my mother tongue was Punjabi, she said that her brother can speak a number of Asian languages such as Punjabi and Hindi, as well as Bengali. Another student told me that his mother is from Tanzania and he can speak some Swahili words whilst another student had learnt some Portuguese. This mathematical activity can generate classroom conversations relating to students' own cultural backgrounds, aiding motivation and helping students to show a greater appreciation of each other. One pre-service teacher commented: 'Helps to get pupils talking about their different cultures.' Mentors and pre-service teachers can utilise the diversity of their students, tapping into the variety of written and spoken languages, to create rich mathematical work. Cantle (2001: 30) states "Many of those we spoke to preferred integration on many levels and those who experienced schools in a mixture of faiths, races and cultures were very positive about that environment."

The second activity is based on the Vedic Square, and Shan and Bailey's book gives a background to Vedic mathematics and the Vedas. The Vedic square has its own unique mysteries, whereby numbers form symmetrical shapes, which were thought to be models of the universe. In the Vedic texts and work of Arab mathematicians, the universality of the mathematics language is emphasised.

Keywords to emphasise are based on the shapes generated from the patterns below such as 'pentagon', 'rectangle', 'triangle' as well as 'rotation' and 'reflection'.

Instructions for completing the square in Figures 4.5 are:

1	2	3	4	5	6	7	8	9
2								
3								
4								
5								
6								
7								
8								
9								

1	2	3	4	5	6	7	8	9
2	4	6	8	1	3	5	7	9
3	6	9	3	6	9	3	6	9
4	8	3	7	2	6	1	5	9
5	1	6	2	7	3	8	4	9
6	3	9	6	3	9	6	3	9
7	5	3	1	8	6	4	2	9
8	7	6	5	4	3	2	1	9
9	9	9	9	9	9	9	9	9

Figure 4.5 Vedic Square

i Complete the matrix by multiplying the horizontal row with the vertical column

ii If the product is greater than 9, add the two digits together, for example, $8 \times 5 = 40 \rightarrow 4 + 0 = 4$. Continue until you get a single digit.

iii Join up all the one's in one colour; join up the two's on another grid; join up the threes on another grid and so on.

iv Pose: "What patterns do you notice?" "What is the sum of each row . . . of each column?" "Number sentences include triples, for example, 3 6 9 $\rightarrow$ 3 + 6 = 9 and 7 5 3 $\rightarrow$ 7 + 5 = 12 $\rightarrow$ 1 + 2 = 3; can you spot any such triples that make a mathematical statement using the four basic operations?"

(Shan & Bailey 1994: 154)

Pre-service teachers stated:

> I really enjoyed the session. I loved the vedic [sic] squares activity, fascinating stuff. It's the kind of activity where I would appreciate guidance on how to make it rich and well structured, so I don't risk prompting general vague discussions about what 2 and 7 have in common etc.
>
> I also liked the patterns (shapes, translations etc.) in the vedic squares – an unusual link between numbers and shapes which demonstrate some of the beauty of mathematics.
>
> The Vedic magic squares. It was quite new and different for me, and also touches on a variety of mathematical topics in a way that invites exploration.

Examples of squaring numbers, squares and square roots and spirolaterals in the Vedic square (all from Shan and Bailey's book) have also proved irresistible for pre-service teachers to trial!

The third activity utilises centrefolds from the publication *New Internationalist* based on statistics from around the world. This resource brings together citizenship issues and challenges stereotypes and pre-conceived ideas about developing nations. These global figures are based on the theme of the magazine for that edition and examples that I have used include:

1 Europe – the facts
2 Patents on life
3 Latin America and the Caribbean
4 Child Labour
5 Judeophobia
6 The other America
7 Bangladesh

Pre-service teachers working together on planning activities based on the above came up with the following ideas, including that of equity, respectively:

1 closed questions based on the unemployment rates for minority groups; open questions on the youth unemployment index
2 representation of graphs in different forms; sizes of countries
3 gap of the rich and poor; debt percentage increase and decrease
4 transforming information in pie charts to tabular format
5 genocide; anti-Semitic incidents; interpretation of data in bar charts
6 inequality and poverty; prediction and projection; prison data
7 differences of aid; economy; percentage of arms; televisions and telephones per thousand.

One pre-service teacher commented: '. . . an excellent cross-curricular activity'. Such statistical activity allows for group work with students in the classroom. Pre-service teachers can share ideas and discuss how the statistics might be used in planning a lesson and identifying the age and level of attainment for which they are suitable: "Can you represent the figures and diagrams in a different chart?" "Can you draw a box-plot for the data in the graph?" "Write the numbers in words". This can lead to presentations of their work from students.

In general, pre-service teachers were very positive about historical and cross-cultural aspects and multicultural mathematics, stating that these were useful activities that could be used in the classroom. The activities were described as purposeful with a lot of scope for differentiation, investigation and collaboration. Pre-service teachers felt this was an area their students would engage with and that they enjoyed this interesting session of original and different approaches to mathematics.

Whilst on school placement, pre-service teachers were asked to plan and teach a lesson for lower secondary-age students on historical and cross-cultural roots of mathematics. As part of becoming a reflective practitioner, they wrote an evaluation of their lesson. Some of the ideas they used included:

1 the seven bridges of Königsburg problem with students' presentations for classroom display
2 Pythagoras' Secret Society; Pythagorean Triples
3 Roman numerals as a number system (Romans, history and Rome)
4 binary numbers
5 research a famous historical mathematician, summarise theories, create a poster and present to group
6 irrational Pythagoreans – Hippassus
7 Leonardo de Vinci's Vitruvian man and class data on lengths and heights
8 problems from Rhind papyrus and Diophantus' gravestone; French duelling
9 write and speak numbers in different languages; Vedic square
10 Archimedes and the Eureka moment – students produce a presentation of quiz questions
11 Chinese method for long division and long multiplication

12 bridges of Königsburg problem – history and patterns in number of even and odd problems; mathematician Euler
13 Fibonacci; Pascal's triangle and its Chinese origins; Golden ratio and Divinity; Sierpinski
14 Sieve of Eratosthenes; multiples and primes; Cuisenaire factors
15 M.C. Escher and tessellations; symmetry and rotation
16 examples from 'The Man who Counted' – an Arabian story to teach fractions in a problem-solving context (Tahan 1993).

For further work on geometry, Roman mosaics and floor tiles can be copied and explored to create new designs by students (see *Geometric Patterns from Roman Mosaics and How to Draw Them* by Robert Field, 2005); for work on transformations (reflection, rotation and translation), students can study Hindu and Arabian architecture, Islamic Art and the African patterns (*Africa Counts: Number and Patten in African Culture* by Claudia Zaslavsky, 1990); drawing symbols of world religions practices compass work; symmetry can be investigated through Celtic designs. These ideas can help students' understand the mathematical concepts and by using technology such as visualizers, teachers can clearly highlight shape properties to students, perhaps at the beginning of a lesson. Students' work can be laminated and creatively displayed to aid dialogue amongst teachers and students.

Activities from the Far East also offer rich opportunities in mathematics. There is a plethora of inspiring classroom activities based on Chinese tangrams, for example, where classroom experience highlights how students with strong visual-spatial awareness excel. Origami also provides opportunities for mathematics learning. In one lesson, a boy stated how, in some countries, origami skills are taught in lessons and I talked about the Japanese art of paper folding. Students all wanted to have a go at producing origami shapes rather than work from the textbook! Japanese culture has also given us the leisure activity of Sudoku, used by mathematics teachers for a starter or homework. Consider also using Kumon activities for mental arithmetic.

An activity used by an experienced mentor, working in a large mathematics department of a secondary school, is to ask pre-service teachers on teaching practice to use the scheme-of-work for one of the lower-secondary school age classes and complete a spiritual, moral, social and cultural audit. This highlighted how mathematics covers all four of these aspects. The mentor had completed an audit as well and was able to compare her own findings with those of the pre-service teachers. This demonstrated that pre-service teachers are able to recognise where mathematics makes such contributions and can include them in their teaching.

This mentor has been embedding historical and cross-cultural aspects in mathematics into the mathematics department schemes of work. One successful classroom activity completed by the mentor includes an Amish quilt discussion with the students during which they are asked if they have any knowledge of the

Amish people. Before the lesson, students compile a presentation showing how the Amish community travel, dress, their housing, religious events, education and the activities they complete due to lack of television, including quilt-making. The quilts can be used to show geometrical shapes and symmetry. The numerical aspects relate to the use of fractions to compare the different colours and shapes used. Students can then design their own quilt using the information gained at the start of the lesson.

A second classroom activity involves using a search engine where students can be asked to type in the objective of the lesson being taught and find a picture associated with it. This is then used to raise spiritual, moral, social and cultural issues. For example, 'Pythagoras' theorem' may generate a picture of Pythagoras and the construction of triangles, leading to Lancelot 'Capability' Brown. An investigation of the use of sticks and string to draw mathematical shapes in the 18th century can be explored and how Capability Brown's work is embedded in the English garden culture.

Discussions amongst teachers in her department suggested working with students late in the school year following examinations to carry out investigations on historical and cross-cultural aspects. Their resource 'Murder in the mathematics department' is based on ciphers and codes used by historical figures such as Julius Caesar. Links can be made to 'The Enigma Project' and the World War II machine.

For topics of weighing, measuring and ratio, consider using recipes from different countries, bringing fruit and vegetables into the classroom; currencies from different countries can aid understanding of exchange rates; equivalent fractions can be explored through the Islamic Law of Inheritance; use local, national and world maps, including students' own travels, to teach scale, distance and proportion. Other examples include investigating the Peruvian number system; in *The Story of Maths*, a series of four BBC television programmes to use with students, Marcus du Sautoy (2008) looks at the contributions of the Greeks, and mathematics in the East, Europe and from different times.

In addition, cultural activities can be explored through cross-curricular opportunities with colleagues, such as during the Football World Cup and Olympics. Many sports have campaigns and projects that promote anti-racism and equality and fight against discrimination and exclusion. Further, mentors can utilise local organisations such as NEAD (Norfolk Education and Action for Development) which work for global equality and justice, addressing racism in schools through an outreach coordinator. Such organisations can offer a range of classroom activities for group discussion and useful resources for colleagues such as subtle racism statements.

An exploration of the issues surrounding mathematics, diversity, racism and the spiritual, moral, social and cultural agenda may be covered, through continuing professional development. This might include working in partnership with primary mathematics mentors and pre-service teachers to help to increase student confidence and self-esteem. Colleagues can be inspired by how essential mathematics is to understanding our diverse world through a range of in-service training interactive topics and challenges such as: using mathematics to encourage active global citizenship; cultural problem-solving tasks; diversity and social justice; wants and needs; financial responsibility; strategy games from around the world; why does the global dimension matter to mathematics? (NEAD Primary 2012).

Pre-service teachers can use ideas from their 'History of Mathematics' modules from undergraduate degree courses. Those from black and minority ethnic (BME) backgrounds and different cultures can enrich their peers' learning by sharing their own personal educational experiences, especially if from other countries. Other individuals can be supported by consideration of their cultural needs, such as being given the use of a prayer room. The discussion of equity and diversity issues and challenging racism amongst pre-service teachers is an important part of learning to teach mathematics in a multicultural society.

Final remarks

Beginning teachers should embrace a multicultural approach to teaching as it enhances the subject and highlights how mathematics can be used to overcome barriers between different cultures. Consider also the positive influence of black and minority ethnic teachers, including those in senior roles, and the numbers of such students across mathematics classes grouped by attainment. On a recent visit to an inner-city multicultural school which staff consider to be a 'true comprehensive', non-white students made up only one-fifth of the mathematics 'top set' in a class of 11–12-year-olds and only one student of the nineteen in a class of 17–18-year-olds; in contrast, half or more of lower sets aged 12–14 were black and minority ethnic students. For all students, we each have a responsibility to address our own preconceptions and assumptions of other cultures and people in society. It is vital that our words and practices reflect beliefs of equality, regardless of nationality or culture. As the great poet Professor Benjamin Zephaniah (1996: 10) states: "For it is written in the great book of multi-culturalism that the curry will blend in with the shepherd's pie and the Afro hairstyle will return."

References

Banks, J.A. (1992) 'Multicultural education: approaches, developments and dimensions', in J. Lynch, C. Modgil, & S. Modgil (eds) *Cultural Diversity and the Schools. Volume 1. Education for Cultural Diversity Convergence and Divergence*, London: The Falmer Press, pp. 83–94.

Cantle Report (2001) *Community Cohesion: A Report of the Independent Review Team*, London: Home Office.

Cockcroft, W.H. (1982) *Mathematics Counts*, London: HMSO.

Critchlow, K. (1976) *Islamic Patterns*, London: Thames and Hudson Limited.

Davis, C. (1989) *Celtic Iron-On Transfer Patterns*, New York: Dover Publications, Inc.

Department for Education (2007) *Mathematics National Curriculum*, London: DfE.

Department for Education and Skills (2007) *Curriculum Review: Diversity and Citizenship*, London: DfES.

du Sautoy, M. (2008) *The Story of Maths*, BBC Television.

Field, R. (2005) *Geometric Patterns from Roman Mosaics and How to Draw Them*, Hertfordshire: Tarquin Publications.

Jagdev, M. (1996) 'Educational Experiences of Black and Asian pupils in an inner-city comprehensive' (unpublished).

Joseph, G.G. (2011) *The Crest of the Peacock Non-European Roots of Mathematics*, 3rd edn, Oxfordshire: Princeton University.

Lawrence, D. (2012) *Doreen Lawrence – Campaigning Women*, BBC Radio 4 Woman's Hour, 31 December 2012. Online. Available at HTTP: <http://www.bbc.co.uk/programmes/p01378ns> (accessed 20 February 2013).

Macpherson Report (1999) *The Stephen Lawrence Enquiry*, London: The Stationery Office.

Modood, T. (2007) *Multiculturalism A Civic Idea*, Cambridge: Polity Press.

New Internationalist. Online. Available at HTTP: <http://www.newint.org/issues/2012/09/01/> (accessed 27 September 2012).

Norfolk Education & Action for Development. Online. Available at HTTP: <www.NEAD.org.uk/Schools> (accessed 27 September 2012).

Palin, M. (2005) *Himalaya*, London: Orion Books Limited.

Parekh, B. (2000) *The Future of Multi-Ethnic Britain The Parekh Report*, London: Profile Books.

Saxe, G. and Esmonde, I. (2005) 'Studying cognition in flux: A historical treatment of *Fu* in the shifting structure of Oksapmin mathematics', *Mind, Culture & Society*, 12 (3 & 4), 171–225.

Shan, S. and Bailey, P. (1994) *Multiple Factors – Classroom Mathematics for Equality and Justice*, Stoke on Trent: Trentham Books Limited.

Tahan M. (1993) *The Man who Counted: A Collection of Mathematical Adventures*, New York: W. W. Norton & Company Inc.

Ulverston Victoria High School (2005) *Maths2Art*. Online. Available at HTTP: <http://www.maths2art.co.uk/> (accessed 10 January 2013).

Vallely, P. (2012) *Race in Britain. Independent on Sunday*, 8 January 2012. Online. Available HTTP: <http://www.independent.co.uk/news/uk/home-news/race-in-britain-2012-has-life-changed-for-ethnic-minorities-6286786.html?origin=internal Search> (accessed 12 November 2012).

Wilkins, V.A. *The Life of Stephen Lawrence*, London: Tamarind Books.

Wright, P. (2004) *Human Rights in the Curriculum: Mathematics*, London: Amnesty International UK.

Zaslavsky, C. (1990) *Africa Counts: Number and Pattern in African Cultures*, New York: Lawrence Hill Books.

Zaslavsky, C. (1995) *Mathematics from Many Cultures*, Australia: Mimosa Publications Pty Limited.

Zephaniah, B. (1996) 'I have a scheme', in *Propa Propaganda*, London: Bloodaxe Books.

Chapter 5

Inclusive approaches to learning and teaching mathematics

Geoff Tennant and Catherine Foley

Inclusion is the principle that, whenever possible, all children should be in the same classroom, forming meaningful peer group relationships, and finding success and challenge in the curriculum. Implementing this principle presents huge challenges to pre-service teachers of mathematics as each brings their own diverse experience of the subject and, in some cases, these experiences may have negative consequences in the classroom. This chapter considers conflicting ideals within inclusion and how to work with diverse groups of children, suggesting activities mentors and pre-service teachers can engage with as they learn to engage all children simultaneously and positively.

Introduction

We start with the premise that a good understanding of mathematics is necessary for all in the workplace and in everyday life and that all students need maximal access to this crucial subject, with interesting, challenging and well-differentiated lessons meeting their needs. Practice-based mentors are vital in supporting pre-service teachers as they learn to work effectively with all students in the mathematics classroom. Strong mathematicians need to understand difficulties faced by students, and pre-service teachers who do not yet feel confident need to be able to communicate an enjoyment of the subject. All pre-service teachers need to be comfortable working in an open-ended manner, creating an environment where it is safe to make mistakes.

Two contrasting pre-service teachers

Consider the following scenarios:

Box 5.1

In one classroom, 10-year-olds are exploring the properties of quadrilaterals. They have chosen from a wide range of equipment, and are focused on exploring what can be created using four straight pieces of equipment in combinations of lengths. The classroom doors have been thrown open, and a pair of children have taken a set of pieces of cane outside and are attempting to find pairs of the same length, muttering to themselves 'now, if we put the two that are the same length together at one corner, what can we make?' Across the room, a small group think they have discovered that if they have three pieces the same length, they can't make a quadrilateral without the fourth piece also being the same length, and are trying to work out whether they are right before taking their discovery to the teacher. At times, learners refer to the multi-language display they made earlier in the week giving key terms in pictures and various classroom languages. All of the class is engaged purposefully, and the teacher has space to observe, mentally noting those who have and haven't mastered the language and key ideas.

In the parallel class, the intended learning is the same but the atmosphere is very different. Children are all seated at their desks, working through a paper-based task to identify, classify and label a series of quadrilaterals. One throws down his pencil in frustration – 'I can't do this Miss!' Others, bored, are starting to murmur to each other under their breath, although they are soon quelled by a look from the front. Separately, a teaching assistant works with a group, supervising them drawing around a series of plastic shapes, colouring in and labelling their properties. She is overheard prompting: 'What is this . . . that's right, it's a rectangle isn't it . . . so we need to label 4 sides'. Some individuals, bewildered by the mathematical language, have resorted to trying to make sense of the task in their home language, but are told that they need to learn the 'English names'.

These two composite examples are based on extensive experience of working with pre-service teachers. Both of these classes are being taught by pre-service teachers on a post-graduate primary initial teacher education course. They both entered the year with what Hobden and Mitchell (2011) characterise as a breakdown in their own mathematical lives, learned helplessness built up after repeated failure. However, their experiences in school have been very different – and that is due to their mentor. The first pre-service teacher has learnt how to slow down, observe the class and get to know them as individuals. She draws on what she knows about how students learn best, and plans accordingly. Her starting point is aiming to include all of the learners – yet she has the confidence to take risks, knowing that her approach won't always work for every learner in the class. She is still insecure about her mathematical subject knowledge, but works hard with the mentor to stop this being a barrier to planning open-ended activities for the learners. The second pre-service teacher has stayed in the mindset developed during her own schooling, that 'washed in' through her own observations of mathematics teaching (Ball, Lubienski & Mewborn 2001: 437). For her, inclusion is about deciding on a set task and then providing differentiated versions, drawing on the support of a teaching assistant for the least 'able' group. She is frustrated by the fact that the history of her own negative experience of mathematics is repeating itself, but can't see past her need to deliver the next learning objective on the list.

These examples serve to illustrate a key difference in developing inclusive practice within mathematics teaching – the role of the mentor. The purpose of this chapter is to explore how pre-service teachers, no matter where their starting point, can be supported to gain confidence in including all learners in their mathematics teaching.

At the heart of inclusive practice is the principle that all learners need a good understanding of mathematics in order to live rich and fulfilling lives. All learners deserve maximal access to this crucial subject, with interesting, challenging and well-differentiated lessons meeting their needs. Learners bring to the mathematics classroom different previous experiences, levels of attainment and attitudes. In tandem, pre-service teachers bring their own hugely varying experiences of the subject, and indeed, views on what the subject is about and who it is for.

Mentors have a pivotal role in working with pre-service teachers, taking them from where they are and helping to equip them to be inspirational teachers across the diverse range of attainment and need presented by the learners in front of them. This chapter aims to address how inclusion can be defined in a mathematics classroom context, and to suggest a vision of an inclusive classroom. Challenges faced by different pre-service teachers are considered, alongside how the mentor can work with such teachers in a meaningful way.

The meaning of inclusion

A useful starting point for a definition of inclusion, albeit specifically within the context of special educational needs, is given in DfES (2001: 2):

- inclusion is a process by which schools, local education authorities and others develop their cultures, policies and practices to include pupils;
- with the right training, strategies and support nearly all children with special educational needs can be successfully included in mainstream education;
- an inclusive education service offers excellence and choice and incorporates the views of parents and children;

- the interests of all pupils must be safeguarded;
- schools, local education authorities and others should actively seek to remove barriers to learning and participation;
- all children should have access to an appropriate education that affords them the opportunity to achieve their personal potential;
- mainstream education will not always be right for every child all of the time.

Equally, just because mainstream education may not be right at a particular stage it does not prevent the child from being included successfully at a later stage.

With the possible exception of children with severe mental and physical disabilities, then, the assumption is that pre-service teachers need to be ready to work effectively with all children, irrespective of factors such as gender, ethnic background, social class, prior learning experiences, home backgrounds, preferred learning style, learning difficulties and current attainment level. Whilst schools may operate setting, streaming or other grouping methods, pre-service teachers need to be ready to teach simultaneously diverse learners with different attainments and attitudes.

The nature of inclusive schools and what it means to be inclusive has attracted considerable attention over the last 30 years or so. One theme which emerges is that simply placing diverse children, including those with moderate learning and physical disabilities, in the same classroom is not enough, and gives rise to the possibility of 'maindumping' (Stainback & Stainback 1990: 20) – where all children are 'dumped' into mainstream schooling with no further thought as to their needs. It is also essential to ensure social inclusion, with all children able to access meaningful peer group relationships (see, for example, Farrell 1997).

The concept of inclusion also encompasses the need for all students to be academically included in the classroom – to experience a curriculum which facilitates success and challenge, motivation and interest. Different views on how this is achieved can be found across the world. Pacific Rim countries, which typically perform well on international league tables, look to work with children in a restricted range of attainment (see, for example, Reynolds & Farrell 1996). An alternative, albeit extreme view, is taken by Stainback and Stainback (1990: 10–11) in advocating that all children should be working according to a common theme, even if this means that when considering temperature, "another student may simply be requested to grasp and hold a corner of the map". In the context of a primary literacy lesson, a report by the English schools inspectorate (Ofsted 2004) rejected this view, advocating an approach whereby children are following a curriculum appropriate to them rather than trying to follow a scheme appropriate to the majority.

It is beyond the remit of this chapter to debate the relative merits of these positions. It is important to note, however, that somewhat different views are held on what represents an 'ideal' within the inclusion debate, with potential conflicts between pursuing an inclusion agenda and other agendas, particularly standards. This points to mentors ensuring clarity as to the school position, in order that pre-service teachers can be given a clear sense of what is expected and how they need to develop.

Ensuring inclusion

In ensuring that all learners are included in the classroom, physically, socially and academically, there are many considerations. Learners should be included irrespective of:

- gender;
- ethnicity;
- language proficiency;
- learning difficulties;
- home background;
- previous experiences of the subject;
- enthusiasm for the subject/perceptions of usefulness;
- current attainment within the subject;
- physical disability;
- vulnerability to bullying for whatever reason.

Again, detailed consideration of each of these areas is beyond the scope of this chapter. Here, the main focus is to consider academic inclusion and how pre-service teachers can be guided towards ensuring good access to mathematics for all children within their care.

What inclusion looks like

Building on the thoughts above, and also on the work of Sharp (2004) and Tomlinson (1999), the key characteristics of an inclusive classroom include:

- fun and engagement;
- children working purposively at an appropriate level on tasks which give opportunities for challenge and success;
- a sense of awe and wonder;
- opportunities for open-ended activities;
- a sense of the connections between different types of mathematics;
- a sense of understanding the usefulness of mathematics;
- opportunities for pupils to work together.

Central to the above is a strong sense of differentiation, ensuring that all children are able to access the curriculum in a manner which ensures success, challenge, motivation and interest. As with inclusion more generally, it is important to note that different people have divergent views as to what represents the goal with differentiation. At the extremes, one ideal is unrestricted differentiation, allowing students to work at their own level and speed, for example through individualised schemes. At the opposite extreme is the idea of restricted differentiation, expecting students to be working within relatively narrow bands within the same classroom.

Although frequently seen as being about planning for different levels of attainment, we could more usefully interpret the term 'differentiation' in terms of doing things differently in order to meet varying learning needs. In principle, differentiation in mathematics lessons might be:

- by quantity: faster working children get through more work;
- by quality: some children produce 'better' work than others;
- by task: different children are given varying tasks;
- through groupwork: a carefully chosen group of children are given a joint task to do, with different children contributing different ideas;
- through peer teaching: getting children who understand something to explain to those who don't;
- through scaffolding: to have a fixed point for the end of the lesson, for example: to draw a pie chart starting with a frequency table with total frequency adding to a factor of 360, giving different children different levels of scaffolding in doing so;
- through support: offering additional help, possibly through the use of teaching assistants or apparatus, to some children to achieve the same outcome;
- through context: enabling different children to explore the same concept in different ways, for example in the indoor or outdoor environment.

If there is no planning for differentiation the first two forms will happen automatically. One criticism made of other forms of differentiation is that they can become 'self-fulfilling prophecies'; once a teacher has decided that a learner is high or low attaining or works in a certain way, the teacher can then treat the learner accordingly, so reinforcing this perception (see, for example, Watson 2006). One possible solution here is to give learners the choice as to which tasks to do or roles to perform – but if they have already decided that they are not very good at mathematics, there is the clear possibility that they will choose to work well within their level of attainment. In practice there needs to be a variety of approaches, with no single one providing all the answers.

The discussion above serves to underline the fact that entering teaching is a complex business, with a considerable amount to take on board. In an American science context, Davis, Petish and Smithey (2006) expressed the view that pre-service teachers start with sophisticated ideas about teaching but without the ability to implement these ideas in a meaningful and sustainable manner. They further considered that, left to themselves, pre-service teachers find ways of coping in the classroom which may not be very helpful, in much the same way that someone left to work out for themselves how to swim across a swimming pool will find some way of doing it, but not necessarily a very efficient manner, being cut off from generations of accumulated experience. The role of the mentor is, therefore, central in guiding and enabling pre-service teachers to translate principles of inclusion into a meaningful reality within the context of the placement school (see, for example, Brookes 2003).

The profiles of pre-service teachers

Pre-service teachers bring with them backgrounds, skills and attributes as diverse as the students in their care. They might have grown up with an implicit assumption that mathematics is a male-friendly subject, unwittingly setting examples and problems in contexts likely to alienate girls, or relying upon a competitive style of engagement, alienating the less confident. They might come from a cultural context in which mathematics is valued and emphasised from an early age, and find it difficult to deal with a lack of basic computational skills, knowledge of number facts or parental support amongst some of their students. Or they might be going into teaching as an undergraduate student with an arts-based specialism, unable to see the purpose and relevance of mathematics to children's lives. The mentor needs to be aware of the range of possibilities, and choose their strategies accordingly. Two particular cases are explored below.

Good mathematicians needing to understand children's difficulties

Strong mathematicians coming into teaching will have been in classrooms where they found mathematics relatively straightforward to learn, with the intrinsic satisfaction that this brings and the sense of being seen to do well. Compounding this is Lortie's (1975) notion of the 'apprenticeship of observation'; virtually all pre-service teachers have spent a large amount of time observing teachers in operation, contrasting with other professions – law, accountancy, medicine, engineering – where entrants will, in the normal course of things, have seen very little or none of the work of serving professionals. This means that entrants to teaching bring considerable preconceptions as to what teaching looks like (see Haggarty 2002). In some contexts, the importance of good subject knowledge for teachers, evidenced by traditional measures of personal success in the subject, is overplayed. So, for example, in criticising the proportion – less than 2% – of primary teachers entering the profession in the UK with a mathematics specialism, The Royal Society (2010) make an implicit claim that quality of mathematics teaching and mathematical qualifications are linked, although evidence for this supposition is surprisingly elusive (Ball *et al.* 2001).

If, in the experience of pre-service teachers with strong mathematical backgrounds, and that of their immediate peer group, mathematics is straightforward, easily understandable and fun, they may have problems understanding children who do not find mathematics as easy – and as fascinating – as they do. These pre-service teachers can struggle to understand the problems faced by many of the youngsters they will be teaching, specifically with unpicking understandings they have gained, often intuitively, over many years. For example, consider the standard equation of the straight line $y = mx + c$ in which, for any one straight line, m and c are constants and x and y are variables. For a strong mathematician, the distinction between constants and variables is so obvious that further explanation is not required. As noted by French (2002) and many others, however, this distinction causes endless problems in the early stages of learning algebra; in many respects less capable mathematicians are at an advantage here in helping students come to their own understandings.

Good general qualities, wary of mathematics

In contrast to this first group, teachers of primary school mathematics may have qualifications some way distant from mathematics, and hold negative views towards the subject. This can easily be communicated, with pre-service teachers having poor subject knowledge likely to retreat towards didactic forms of teaching, particularly if that is how they were taught at school themselves. Such teachers may have developed the coping strategy of avoiding mathematical tasks themselves in any kind of public situation (Bibby 2002), and the mentor needs to be sensitive to the personal as well as professional journey in store for these pre-service teachers.

An evidence-based approach may help here; for example, knowing that female teachers with a high degree of mathematics anxiety are likely to transfer this to their female students, resulting in a perpetuation of gender stereotypes about mathematics (Beilock, Gunderson, Ramirez & Levine 2010), can be a first step in encouraging pre-service teachers to consider the inclusive role models they need to provide for their students.

Another strategy for this group is to focus their attention less on mathematical subject knowledge, and more on how they can transform this knowledge into a form that is pedagogically useful and accessible for children. This can include mathematical representations in terms of analogies, illustrations and images, explanations and demonstrations (Rowland *et al.* 2009). In the context of inclusion, this choice of examples and analogies can be a key strategy for overcoming barriers to learning. For example, drawing on the strong oral tradition of a group of Romany students by setting a calculation in a real-life narrative can emphasise their strengths, rather than relying on more problematic starting points such as symbolic notation (Stathopoulou & Kalabasis 2007).

Strategies for mentors

Up to this point, we have set out some of the opportunities and issues that the mentor and pre-service teacher need to be aware of in relation to inclusion and the teaching of mathematics. Expectations on teachers to deal with the wide range of students' needs are greater than ever, with pre-service teachers expected to engage with issues of a level of complexity unimaginable a generation before. The rest of this chapter suggests how mentors can support their pre-service teacher in meeting these demands.

Be clear about your school's position on inclusion and differentiation

As noted above, within the inclusion and differentiation debates there are different positions to be held. So questions to be addressed include:

- how is inclusion understood within your school? Is there a policy that the pre-service teacher needs to be aware of?
- what difference does an understanding of inclusion make in practice? Can you introduce the pre-service teacher to a student who has been successfully included within mathematics lessons as a result of careful planning and targeted provision?
- what elements of your teaching would you consider to be influenced by the inclusion debate?
- how is differentiation understood within your school?
- how is differentiation operationalised? Are there any models that the pre-service teacher could observe with different teachers or within other subjects to deepen their understanding? How does this impact on learners?

Get to know the pre-service teacher

Hobson *et al.* (2009) emphasised the importance of mentors getting to know their pre-service teachers, not least in guiding them to work in inclusive ways. Indeed, this is very much within the spirit of a key part of the inclusion agenda; get to know your learners! As noted above, different pre-service teachers will bring very different agendas to their teacher training, with diverse training needs – and in some cases, the capacity to take the lead on specific developments within the school. It is worth noting that mentors can be perceived as having unduly high expectations of mature pre-service teachers (Butcher 2000), not least those of minority ethnic background (Devereux 2001). Williams *et al.* (1998) found that different mentors offered different learning experiences for the pre-service teacher, which perhaps one might reasonably expect. If mentors are to take pre-service teachers from where they are and move them on, they will need to be aware of a variety of ways of carrying out the mentor role, perhaps through discussions with other mentors. This approach is also addressed in Chapter 6.

Model good inclusive practice

As noted above, there are different views as to what inclusive practice might mean, so it is important to consider the immediate school context and to ensure that pre-service teachers are seeing clear models. There are reasons why pre-service teachers need to be planning lessons at a level of thoroughness beyond experienced teachers; it is rather more difficult to explain why they should be adopting practices within differentiation and other aspects of the inclusion agenda which experienced teachers are not!

One opportunity might be exploring with the pre-service teacher the provision made for students with English as an Additional Language (EAL). It is well documented that whilst students' 'basic interpersonal communication skills' develop rapidly, their 'cognitive academic language proficiency' lags behind, in some cases taking several years to fully develop (Cummins 2000). This can make it easy for the pre-service teacher to miss the need to plan specifically to include these learners. Making explicit how students are encouraged to explore concepts in their first language, using dual-language dictionaries, modelling the use of everyday language and visual images, and moving from concrete to abstract will help the pre-service teacher to understand how inclusivity can be put into practice.

Arrange small-scale focused observations

Observation is in itself a skill. If an experienced teacher is working well with a class, there may be all kinds of things going on beneath the surface which help to build an inclusive classroom atmosphere but are difficult to observe. This is partly because pre-service teachers need to infer what might have happened in different circumstances. With a teacher and 30 children in a room, and possibly classroom assistants, there are many possible foci for observation. A few examples are given below:

- observing a learner or small number of learners: what are they doing on a minute-by-minute basis?
- observing the teacher: how are transitions managed? What examples are used to introduce a new topic?
- what language is used, both in whole-class discussions and when learners are talking informally amongst themselves?

For strong mathematicians, it is helpful for them to observe learners working on mathematics for whom the subject is difficult to access. Mentors and pre-service teachers need to agree the questions to address through the observation, which may include:

- what language does the learner use to talk about mathematics?
- what understandings are learners bringing to the tasks in hand? Are links made with their own experience?

- what exactly is it that learners are finding difficult? Is it the language of the task, or the underlying logic?
- where mistakes are made, is it possible to infer where the mistake may come from? Are they based on an overgeneralisation, incorrect use of symbols, or a misunderstanding of technical mathematical vocabulary?
- what would be a possible course of action in dealing with these problems?

For less confident mathematicians, other observations may be helpful, including close observation of learners engaging with and enjoying their work. Again, it is helpful to agree with the pre-service teacher questions to address whilst observing, including:

- exactly what are the learners doing?
- what language do they use to speak about the mathematics?
- what lines of logic are they following?
- how are mistakes dealt with by the learners themselves? By the teacher?

Ask the pre-service teacher to work with a small group

Following observations, it is helpful for the mentor and the pre-service teacher to discuss what it is that has been learnt and how ideas can be taken forward. It is then valuable to have pre-service teachers planning to teach mathematics to small groups. Removing the focus on whole-class management can give pre-service teachers an opportunity to focus on overcoming barriers to learning for particular learners and learning objectives.

Encourage the pre-service teacher to talk and listen to learners

Learners can provide the greatest insight into whether they are being included in their mathematics lesson, although their views are sought all too rarely. In theory, one of the advantages of being a pre-service teacher is having the opportunity to gather learners' views and listen to their opinions. It is vital that they be supported to use this time constructively, rather than reacting defensively. We know that when learners feel invisible or overlooked in their mathematics lessons, they begin to lose trust in their teacher (Bibby 2008). Innovative approaches to gaining the learner's perspective, such as those employed by Walls (2007) and Borthwick (2011) in asking learners to draw pictures of themselves in mathematics lessons, can prove hugely useful in gaining a non-language-based insight into learners' views on mathematics teaching.

Encourage pre-service teachers to assess and build on students' understanding

Following Black and Wiliam's literature review on the uses of assessment to aid learning (1998a) much has been written about assessment for learning (AfL) (see, for example, Black & Wiliam 1998b; Black, Harrison, Lee, Marshall & Wiliam 2002; Lee 2006). Whilst concern has been expressed that teachers can end up 'going through the motions' of techniques associated with AfL rather than really getting to the roots of what is happening (Boyle & Charles 2010), the principle of using assessment evidence as a means of guiding children's learning is an important part of developing inclusive practice in pre-service teachers.

At a very informal level, teachers are assessing learners by asking questions, noting the answers and then following up with further questions or other interactions based on what the learner has said. Alongside modelling good practice for pre-service teachers is the need to ensure that they are explicitly aware of the nature of the interchanges that have taken place and how they might differ according to how learners respond.

Assessment also is undertaken through marking of tests, homework and classwork. Orton (2004) gives an example from his own experience as a primary school learner doing a mathematics test each week with slightly differing numbers, repeatedly getting a question relating to finding the difference wrong. Whilst he was told his answers were incorrect, no support was given until eventually another learner told him what was going on, leading to the question: what was the purpose of this assessment? Here, the pre-service teacher benefits from clear guidance as to how to make good use of their time in marking to ensure that they are in a better position to guide students' learning. Further questions to consider regarding assessment might be:

- can learners be guided towards self-assessment or peer-assessment?
- might the pre-service teacher, at least occasionally, give learners work to do depending on how they have performed up to that point?
- rather than give a mark, might the teacher make a comment with the first activity of the next lesson being for the learner to respond to that comment?

Look in meetings to move from immediate concerns to more overarching issues

When pre-service teachers start to work in the classroom it is likely that they will need to address very immediate teaching and classroom management issues. Whilst pre-service teachers are working at this basic level of developing the craft of the classroom, it seems to follow that the mentor needs to be wary of making too many other demands.

A number of authors express the view, however, that frequently, discussions between mentor and pre-service teacher do not progress from this initial level, with Haggarty *et al.* (2011) concerned that too much attention is paid to behaviour management and 'fitting in' to the school ethos rather than engaging with specific issues of teaching and learning.

Allow autonomy and opportunity to try things out

Mentors can sometimes be criticised for being too prescriptive, or for imposing their own model of teaching on their pre-service teacher (see, for example, Geen 2001). For pre-service teachers to be encouraged to be professionals in their own right, there needs to be a sense of allowing them the autonomy to make up their own minds, and, where necessary, to make mistakes. Of course, students' learning needs to be paramount, but this must not be used as a reason to constrain pre-service teachers.

Mentors need, therefore, to encourage the pre-service teacher to think creatively if other strategies aren't working. Sometimes, learners can be included by radically changing the approach or context of the typical mathematics lesson. A key difference between the expert and pre-service teacher is the repertoire of previous case notes that the expert can draw upon (Berliner 2002), including unusual strategies that have worked in the past.

So questions for pre-service teachers to consider, with support from their mentor, include:

- would a games-based approach work? The element of chance introduced by using games can be extremely empowering for students, encouraging them to take part in the lesson when otherwise they were finding it difficult to engage. Ideas for this can be found at the NRICH website www.nrich.maths.org;
- could the mathematics lesson be taken outside? Removing the negative context of the usual teaching environment can make all the difference to students needing a fresh approach;
- could groupings be radically altered? Would gender groupings support less confident girls to engage with the mathematics, building their confidence before returning to a more traditional approach? Would a peer buddying system within mathematics provide another route to inclusion?
- does the pre-service teacher have their own skills and passions which could be drawn upon to include and engage students? Perhaps the art specialist working in the primary school could explore the underlying structure within the work of Mondrian or Escher, as suggested in Chapter 4, or the scientist could draw upon their knowledge of the Fibonacci sequence in nature?

Final remarks

For any of these approaches to provide the pre-service teacher with strategies for a more inclusive mathematics classroom, the role of the mentor is crucial. Without the mentor's support to experiment, draw on best practice and try out new ideas, and, above all else, stop and reflect upon the impact on learning, the pre-service teacher is likely to revert to playing safe, unwittingly continuing the cycle of exclusion from mathematical empowerment which has plagued classrooms for far too long.

Despite the fact that engaging pre-service teachers with inclusive practice in the way described within this chapter has been previously unimaginable, the authors' experience working in initial teacher education is that pre-service teachers are able to rise to the challenges presented to them. This chapter sets out a vision of what inclusion means in the mathematics classroom, along with some of the associated debates. Mentors have a central role in modelling good practice and encouraging pre-service teachers to take on board the challenges of inclusion in creative and flexible ways. Preparation of the next generation of teachers will help to ensure more meaningful and enjoyable access to mathematics for all in years to come.

References

Ball, D., Lubienski, S., & Mewborn, D. (2001) 'Research on teaching mathematics: the unsolved problem of teachers' mathematical knowledge', in V. Richardson (ed.), *The Handbook of Research on Teaching*, New York: Macmillan, pp. 433–56.

Beilock, S., Gunderson, E., Ramirez, G., & Levine, S. (2010) 'Female teachers' math anxiety affects girls' math achievement', *Proceedings of the National Academy of Sciences*, 107(5), 1860–3.

Berliner, D. (2002) 'Learning about and learning from expert teachers', *International Journal of Educational Research*, 35, 463–82.

Bibby, T. (2002) 'Shame: an emotional response to doing mathematics as an adult and a teacher', *British Educational Research Journal*, 28(5), 705–21.

Bibby, T. (2008) 'The experience of learning in classrooms: moving beyond Vygotsky', in T. Brown (ed.), *The Psychology of Mathematics Education: a Psychoanalytic Displacement*, Rotterdam: Sense Publishers, pp. 37–61.

Black, P. & Wiliam, D. (1998a) 'Assessment and classroom learning', *Assessment in Education*, 5, 7–74.

Black, P. & Wiliam, D. (1998b) *Inside the Black Box: Raising Standards Through Classroom Assessment*, London: School of Education, King's College.

Black, P., Harrison, C., Lee, C., Marshall, B., & Wiliam, D. (2002) *Working Inside the Black Box: Assessment for Learning the Classroom*, London: Department of Education and Professional Studies, King's College.

Borthwick, A. (2011) 'Children's perceptions of, and attitudes towards, their mathematics lessons', *Proceedings of the British Society for Research into the Learning of Mathematics*, 31(1), 37–42.

Boyle, W. F. & Charles, M. (2010) 'Leading learning through assessment for learning?', *School Leadership and Management*, 30, 285–300.

Brookes, W. (2003) 'GTP training – is adequate good enough?' Paper presented to the British Educational Research Association Annual Conference, Heriot-Watt University, Edinburgh, 11–13 September.

Butcher, J. (2000) 'Subject culture, pedagogy and policy on an open learning PGCE: can the gap be bridged between what students need, and what mentors provide?' Paper presented to the European Conference on Educational Research, Edinburgh, 20–23 September.

Cummins, J. (2000) *Language, Power and Pedagogy*, Clevedon: Multilingual Matters.

Davis, E. A., Petish, D., & Smithey, J. (2006) 'Challenges new science teachers face', *Review of Educational Research*, 76(4), 607–51.

Devereux, J. (2001) 'Ethnic minority students in initial teacher training: issues in support'. Paper presented to the British Educational Research Association Conference, University of Leeds, 13–15 September.

DfES (2001) *Inclusive Schooling: Children with Special Educational Needs*, Nottingham: DfES Publications.

Farrell, P. (1997) 'The integration of children with severe learning difficulties: a review of the recent literature', *Journal of Applied Research in Intellectual Disabilities*, 10, 1–14.

French, D. (2002). *Teaching and Learning Algebra*. London: Continuum.

Geen, A. (2001) 'Debriefing student teachers: improving current practice', *Teacher Development*, 5, 25–38.

Haggarty, L. (2002) 'What does research tell us about how to prepare teachers?' Paper presented at the *ESCalate PGCE conference*, University of Nottingham, 9 July.

Haggarty, L., Postlethwaite, K., Diment, K., & Ellins, J. (2011) 'Improving the learning of newly qualified teachers in the induction year', *British Educational Research Journal*, 37, 935–54.

Hobden, S. & Mitchell, C. (2011) 'Maths and me: using mathematics autobiographies to gain insight into the breakdown of mathematics learning', *Education as Change*, 15(1), 33–46.

Hobson, A. J., Ashby, P., Malderez, A., & Tomlinson, P. D. (2009) 'Mentoring beginning teachers: What we know and what we don't', *Teaching and Teacher Education*, 25(1), 207–16.

Lee, C. (2006) *Language for Learning Mathematics: Assessment for Learning in Practice*, Maidenhead: Open University Press.

Lortie, D. C. (1975) *Schoolteacher: A Sociological Study*, Chicago: University of Chicago Press.

Ofsted (2004) *Special Education Needs and Disability: Towards Inclusive Schools*, London: HMSO.

Orton, A. (2004) *Learning Mathematics: Issues, Theory and Classroom Practice*, 3rd edn, London: Cassell Education.

Reynolds, D. & Farrell, S. (1996) *Worlds Apart? A Review of International Surveys of Educational Achievement Involving England*, London: HMSO.

Rowland, T., Turner, F., Thwaites, A., & Huckstep, P. (2009) *Developing Primary Mathematics Teaching*, London: Sage Publications Ltd.

Royal Society (2010) *Primary Science and Mathematics Education: Getting the Basics Right*, London: Royal Society.

Sharp, B. (2004) *Meeting SEN in the Curriculum: Maths*, London: David Fulton.

Stainback, S. & Stainback, W. (1990) 'Inclusive schooling', in S. Stainback & W. Stainback (eds) *Support Networks for Inclusive Schooling: Interdependent Integrated Education*, Baltimore: Paul H. Brookes, pp. 3–23.

Stathopoulou, C. & Kalabasis, F. (2007) 'Language and culture in mathematics education: reflections on observing a Romany class in a Greek school', *Educational Studies in Mathematics*, 64(2), 231–8.

Tomlinson, C. A. (1999) *The Differentiated Classroom: Responding to the Needs of all Learners*, 1st edn, Alexandria: Association for Supervision & Curriculum Development.

Walls, F. (2007) '"Doing maths": children talk about their classroom experiences', in J. Watson & K. Beswick (eds), *Mathematics: Essential Research, Essential Practice (volume 2): Proceedings of the 30th Annual Conference of the Mathematics Education Research Group of Australasia*, Wahroonga, Australia: Mathematics Education Research Group of Australasia, pp. 765–74.

Watson, A. (2006) *Raising Achievement in Secondary Mathematics*, Maidenhead: Open University Press.

Williams, E. A., Butt, G. W., Gray, C., Leach, S., Marr, A., & Soares, A. (1998) 'Mentors' use of dialogue within a secondary initial teacher education partnership', *Educational Review*, 50(3), 225–39.

Chapter 6

Developing models of collaborative working

Julie-Ann Edwards and Paul Wilson

Collaboration amongst teachers is a well-established part of most effective teaching contexts. This is no less the case for mentors with pre-service teachers, in either solo, paired or multiple placements. This chapter draws on research evidence to explore some of the collaborative experiences which can be modelled by practice-based mentors for beginning teachers of mathematics to provide them with a grounding in developing similar forms of collaborative practice during their training and in their future teaching careers. We also examine the affordances and constraints of models of practice-based paired placements and address theoretical perspectives which provide mentors with a framework for decision-making in relation to the individual needs of pre-service teachers in these collaborative learning settings.

Introduction

Collaborative activity is a central theme in this book. From co-authorship models to collaborative practices in and outside the classroom, we see this as an essential feature of strong mentoring relationships which help to provide the sort of balance between nurture and challenge essential to the effective development of beginning mathematics teachers. King (2004: 199) claims that, "Many schools have been quick to see that the mentoring process promotes staff engagement with teaching and learning and can lead to school improvement and high pupil achievement". In 1994, Hagger and McIntyre wrote:

> Collaborative teaching, as a mentoring strategy, is any teaching in which the student [pre-service teacher] plays a specified part alongside an experienced teacher. The primary rationale for collaborative teaching is the need for student [pre-service] teachers, in the early stages of their experience, to engage in real teaching without it being too demanding for them. Collaborative teaching enables student [pre-service] teachers to learn thoughtfully in a secure environment and by ***identifying*** with the work of experienced teachers.
>
> (Hagger & McIntyre 1994: 4, emphasis added)

While we endorse this model of collaboration for beginning teachers of mathematics, we caution against such a model being used to promote an apprenticeship model which encompasses the beliefs and attitudes about mathematics held by a mentor. Earlier chapters have alluded to the importance of beliefs about and attitudes to mathematics as key to beginning teachers developing effective practice for mathematics teaching and learning. We add emphasis to 'identifying' in the quote above to illustrate the importance of beginning mathematics teachers establishing their own identity both in relation to mathematics and in the mathematics classroom, particularly through reflection, as suggested in Chapter 12.

A rationale for collaborative activity

A move from a practice-based supervisory role to a mentoring role necessitates a different set of skills in an experienced teacher. Not only does a mentor need sound teaching skills which impact positively on their students, but they also need to overtly demonstrate, with adults, the qualities associated with successful teaching of students, such as effective listening skills, openness to learning-oriented discussion, flexibility, an accommodating approach, accessibility (within reason), a willingness to engage with theory-based approaches to teaching and an empathy with the beginning teacher. This can be seen, by some mentors, as a difficult task when a pre-service teacher appears to be particularly needy. We propose that viewing the mentor–mentee roles as interdependent is likely to enhance the experience for both the pre-service teacher and mentor. This interdependent means of collaboration is supported in research studies such as those by Aderibigbe (2011) and Meirink, Imants, Meijer and Verloop (2010). In particular, Meirink and her colleagues argue that the mutual learning potential of collaboration depends on the level and quality of interdependence within the collaborative relationship. They cite evidence from Little (1990) and Rosenholz (1989) who claim that the highest level of learning between teachers occurs where there are high levels of interdependence. They suggest that, where shared 'stories' are the sole source of collaboration, often informally and in passing, low levels of interdependence are generated. Situations where a teacher assists in another's teaching produces the next level of interdependence. The third level of interdependence depends on more effective 'sharing' of teaching ideas and of resources which impacts on discussions about the curriculum. We suggest that this is a level of interdependence which might be a more mutual experience between a pre-service teacher and their mentor. Indeed, at a mentor training session, one of our practice-based mentors announced:

> Well, they're [pre-service teachers] the ones with all the time to prepare lessons, so it's only right that I expect them to be able to resource our jointly-planned lessons creatively. And, when there's two of them [pre-service teachers in a paired placement] there's twice the value.

Wallis (2012), similarly, records that pre-service teachers bring a range of expertise into a practice-based setting which impacts on wider school effectiveness. The final level of interdependence, identified by Meirink and colleagues, is determined by joint planning of lessons and agreed approaches to problem solving. As a pre-service teacher gains experience, this final level of interdependence is also achievable, as it represents a collective responsibility for the teaching and learning occurring in a learning setting.

Viewed as professional development, the role of the mentor can be 'a mutual support mechanism' (Aderibigbe 2011: 6). For example, Lopez-Real and Kwan (2005) advocate that the communication skills of both the pre-service teacher and the mentor are enhanced in a situation of high interdependence. We contend that developing a mutual language, as soon as possible, through which each understands the other, is essential to facilitate communication about what can prove to be difficult contexts to negotiate, particularly in teaching a complex subject such as mathematics. Ensuring the pre-service teacher is 'on the same wavelength' as the mentor will prove a significant step to mutual interdependence. The ideas associated with such socio-constructivist co-construction of mutual learning are explored further in Chapter 8. Lopez-Real and Kwan further claim that, as the pre-service teacher gains experience in a highly supportive context, there is also the opportunity for mutual learning to occur in joint teaching settings. We now turn to addressing how such 'sharing' might be developed as part of a mentoring context.

Collaborative mentoring activity

The increasing deployment of teaching assistants, in the practice-based setting, provides us with a model for collaboration between adults in lessons. The pre-service teacher can be similarly deployed in the initial stages of their placement, with an expectation of targeted development over a specified period of time. With Hagger and McIntyre's (1994) description as a starting point, we briefly outline some ways in which beginning mathematics teachers and their mentors can effectively collaborate in the teaching setting:

- the pre-service teacher acts as a teaching assistant, with a specific role and focus for observation, in a lesson planned and taught by the mentor;
- the pre-service teacher takes responsibility for the learning of a small group in a lesson planned and taught by the mentor;
- the pre-service teacher and mentor jointly plan a lesson in which the pre-service teacher teaches a specific element;
- the pre-service teacher acts as a teaching assistant, with a specific focus for observation, in a lesson planned by the pre-service teacher but taught by the mentor;
- the pre-service teacher teaches a lesson planned by the mentor who takes the role of a teaching assistant;

- the pre-service teacher and mentor jointly plan a lesson which is taught by the mentor;
- the pre-service teacher and mentor jointly plan a lesson which is taught by the pre-service teacher;
- the pre-service teacher and mentor jointly plan a lesson which is taught jointly by both parties;
- the pre-service teacher collectively plans lessons with a team of teachers, including the mentor, then observes the mentor teaching the lesson;
- the pre-service teacher collectively plans lessons with a team of teachers, personalises this plan, then teaches the lesson;
- lessons are evaluated independently by the pre-service teacher and mentor, then jointly discussed;
- lessons are jointly evaluated.

Underpinning each of these means of collaboration is the rationale for doing any or each of these at a particular point in the pre-service teacher's experience and the specific focus for planning, teaching or observation in these scenarios. Clearly defined reasons for the focus, centred on pedagogical mathematics knowledge and understood by the pre-service teacher, and appropriate timing for enacting each of these scenarios are key to the success of the collaborative activity. For example, the first three of these ways of collaborating can help the pre-service teacher to understand the role of elements of the lesson plan in maintaining the flow of the lesson, managing transitions and enhancing students' learning. Where either the pre-service teacher or mentor takes responsibility for the learning of a small group, more detailed questioning and analysis of students' learning can form the basis of either informal or formal conversations after the lesson. Jointly planning with a team of teachers can help the pre-service teacher to interpret longer-term schemes of work devised by a year team or a mathematics department, to understand how resourcing of these lessons can be adapted for different settings, and how this relates specifically to differentiated provision for individual students' learning. In these co-planning scenarios, whether these are with the mentor or a team of teachers, it is important to remember to make the decisions inherent in *craft knowledge* (Schön 1983) overt for the pre-service teacher, so that they are genuinely acting as a legitimate peripheral participant in a community of practice (Wenger 1998), an idea taken up in Chapter 10.

While we have argued that high levels of mutual interdependence offer the 'best return' for the time and effort required to effectively mentor a pre-service teacher, we do not underestimate the value of informal shared 'stories' in the development of pre-service teachers. Indeed, just such informal conversations are cited in Chapter 8 as 'critical moments' in moving pre-service teachers forward in both their thinking about teaching and their skills in practice. We recognise that these 'stories' can be triggered by a mutually experienced classroom circumstance, a situation experienced by the pre-service teacher which occurred in a classroom other than the mentor's, a query raised by a colleague about the pre-service teacher, or a question asked by the pre-service teacher about written or other requirements by the provider institution, and that these triggers provide an invaluable resource for both informal and more structured conversations related to the pre-service teacher's development. These conversations can be crucial in turning what pre-service teachers often see as personal failings into targeted future development. What is essential for the pre-service teacher is an understanding of how these informally communicated ideas align with aspects of pedagogical teaching knowledge and skills which enable them to address competence-based assessment and to develop *craft knowledge*. The importance of informally generated knowledge amongst pre-service teachers is now addressed in the evidence relating to paired and multiple practice-based placements.

Paired and multiple practice-based placements

The global use of paired and multiple practice-based placements is widespread in the western world. For example, in the United States, professional development schools (PDS) are sites in which the whole school becomes responsible for developing large numbers of pre-service teachers who reflect the geographical diversity of the PDS context. This model is evident also in the training of pre-service teachers in Scandinavia and Eastern Europe. Professional development projects such as the Project for Enhancing Effective Learning (PEEL) in Australia, and the Learning Strategies Group (LSG) in Canada, indicate how pre-service teachers' development is enhanced through practice-based placements within these research-based projects. In England, evidence of how paired or multiple practice-based placements impact on the experience of pre-service teachers, their mentors and their practice-based setting is less rigorously recorded. However, Smith (2004), Sorensen *et al.* (2002) and Sorensen (2004) provide evidence from their studies which point to the optimum conditions in which these practice-based placements can be successful. We also offer evidence from our own research on paired practice-based placements in mathematics departments in secondary schools in England (Wilson & Edwards 2009).

A rationale for paired and multiple placements

Paired and multiple practice-based placements have a varied history in the work of the initial teacher education of pre-service teachers. Motivations for practice-based partners to participate in the training of pre-service teachers, particularly in solo placements, vary widely and include:

- an altruistic expectation to be part of educating future teachers of mathematics;
- an expectation that pre-service teachers can contribute significantly to improved resourcing of lessons, to enhancing schemes of work and to providing ideas for teaching;

- an opportunity to 'cast an eye' over potential employees for the school;
- providing a platform for professional development, through reflection, for staff who are practice-based mentors;
- providing professional development across a wider team of teachers;
- improving the collaboration amongst established teachers in a particular team.

Ofsted (2003: 200) observes that "Schools that participated in ITT [initial teacher training] partnerships almost invariably recognised the invaluable professional development that accrued for teachers and departments when they were involved in training new teachers". However, Wilson (in Wilson & Bolster 2011) recognises the conflict in which some potential practice-based partners find themselves – torn between the political pressures of a government agenda focused on achieving high levels of student performance and supporting the development of new teachers. While we acknowledge these pressures, we argue the case for engaging with paired and multiple practice-based placements for any or all of the factors listed above.

Research evidence on paired placements

Smith's (2004) work on models of pre-service teaching placements and the research of Sorensen *et al.* (2002) and Sorensen (2004), which identified the potential benefits arising from pre-service teachers' paired working, formed the basis for our initial development of paired placements within our practice-based partnerships. Sorensen's work identified three features of successful paired practice-based placements: significant gains were identified for most trainees in terms of emotional and physical security and a sense of belonging; higher levels of achievement in competence-based assessment were most often attained where pre-service teachers shared teaching groups; and those pre-service teachers who had actively prepared for the use of the paired practice-based placements also more often achieved the higher levels of achievement in competence-based assessment.

Smith's model of pre-service placement emphasises that both pre-service teachers should contribute to paired lessons, regardless of which student is leading the class at the time. He uses the analogy of riding a tandem bicycle to illustrate this way of working. The *riders* may alternate and share the responsibility for *steering* at different times, but both must *pedal* together all of the time in order to make good progress. Nokes, Bullough, Egan, Birrell and Merrell Hansen (2008) state that an important finding from their study of paired pre-service teacher placements is the inherent creation of shared experiences, which necessitate conversation and reflection. Such conversations and reflection on shared experiences, they claim, result in "shared investment and shared understandings, both essential elements of teacher development" (2008: 2174). In our research studies, we sought to examine these claims in relation to our respective practice-based partnerships.

In the early stages of our work with paired placements, our practice-based partners experimented with models of paired work with classes. Amongst the approaches used were:

- team teaching;
- alternating the lead over a series of lessons or a topic;
- designating one of the pre-service teachers as the lead teacher and the other as the teaching assistant for all lessons with a particular class;
- alternating the roles of lead teacher and teaching assistant, lesson by lesson;
- solo teaching parallel classes in a year group to enable joint planning;
- solo teaching widely differently-attaining classes of students of the same age, then swapping classes half-way through the practice-based placement, requiring pre-service teachers to exchange significant levels of information about teaching, learning and assessment of each of these classes.

Evidence from evaluative questionnaires and interviews with pre-service teachers, their mentors and senior personnel in the practice-based settings was analysed according to Maslow's (1970) model of a hierarchy of needs (see Figure 6.1). This model structures human needs in three levels: physiological and emotional, intellectual and aesthetic, and meta-cognitive self-realisation. Maslow associated these needs with both the personal and professional lives of individuals. At the

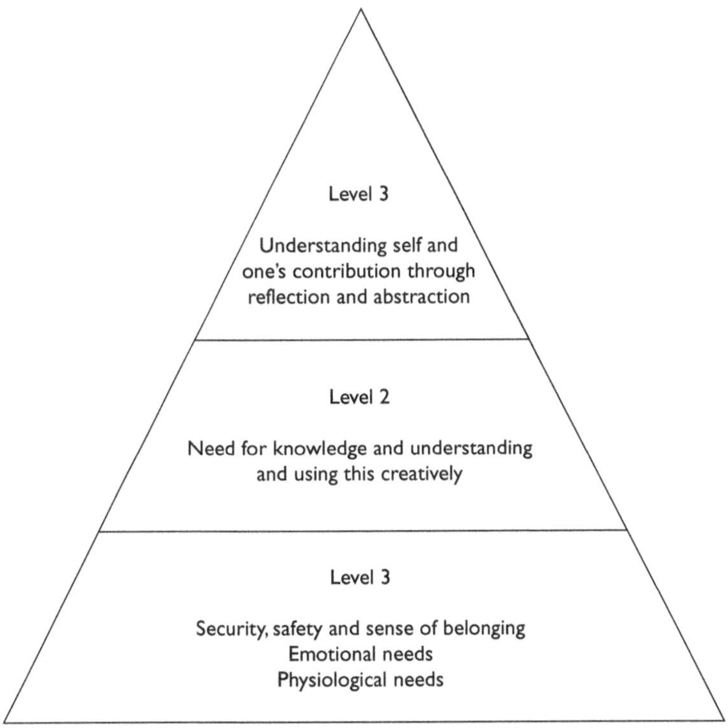

Figure 6.1 Maslow's hierarchy of needs (after Ventegodt, Merrick & Andersen 2003)

lowest level (physiological and emotional), Maslow considered professional aspects to be related to the sense of security and to the individual's need to be acknowledged. At the second level, pre-service teachers' professional needs are more evident in the need for knowledge and understanding, an essential factor for the development of their teaching skills.

Also at level two, aesthetic needs in using knowledge and creative talents in productive endeavour relate directly to the work of teaching. For the highest level, abstraction of the previous needs to understand the self sufficiently to recognise the contribution that the individual is able to play in the immediate, and wider, community lends itself to pre-service teachers' understanding of the quality of self-reflection required to develop effectively as teachers. We contend that this model provides mentors, working with pre-service teachers, especially in paired or multiple placements, with a means of assessing where each pre-service teacher lies in this hierarchy and what they might do to move them further through these levels. Working within this model, we outline evidence from our own practice-based partnerships.

Compatibility of paired students

Evaluations of paired practice-based placements were undertaken in two forms: formally through evaluative questionnaires within the initial teacher education programmes, and more informally through interviews. These identified compatibility of the paired practice-based partners as one of the most important factors in the success of a pairing. This factor appeared to be unrelated to the strength of each student as a potential teacher:

> Our differences in strengths was never an issue because I had a tendency to over-plan at times, so having Greg [pseudonym] to bring me back down to earth was really useful, and Greg had a tendency to under-plan, so I could help out with that, so it swings both ways, I think.

Indeed, a strong pre-service teacher, who reflected on her paired experience with a weaker pre-service teacher (who subsequently withdrew from the initial teacher education programme), indicated that the pairing had forced her to focus her reflective practice more efficiently. She believed that it was this experience which had enabled her to complete the programme with the highest grading against the standards for the award of qualified teacher status (TDA 2008). Differences in teaching styles were also seen as a strength by some:

> I'm very bold and Alison [pseudonym] isn't. I learnt from her style – I'm quite loud and I learnt to bring that down a bit. Where I tackled things one way, Alison tackled things another, so there was variation of styles in the classroom.

In assigning pairings to practice-based partnerships, we rejected the argument that, since pre-service teachers are required to work collaboratively with colleagues as a criterion for competence-based assessment, careful matching of pairings is not an important consideration. We believed that pre-service teachers often feel particularly vulnerable on practice-based placements, and fundamentally more vulnerable than established teachers. The interviews confirmed this assertion, suggesting that security, belongingness and esteem were essential elements for students to begin to establish themselves as developing teachers. One pre-service teacher said that one of the benefits of experiencing a paired placement was having

> someone to share problems in a very similar situation because experienced teachers are very different.

There was unanimous agreement amongst the students that being with someone experiencing the same situations had a positive effect on their engagement with teaching. The term *in the same boat* recurred throughout the interviews, in reference to shared experiences.

One of the factors we took into account, when assigning practice-based pairings, is the logistics of travelling to practice-based partnership sites. We may place two compatible students together, one of whom is able to transport the other to the school. Such pairings reported the benefits of being able to overcome nerves on the journey to school and 'off-load' about disappointments/issues/frustrations on the journey home, describing scenarios that they might not share with teaching colleagues assessing them:

> Just silly things – if you don't feel you've got on with a particular member of staff, just being able to talk that through, or strategies to deal with particular difficult kids.

Several pre-service teachers asserted that this enabled them to concentrate more efficiently on planning for the next day (while sharing the financial burden of travelling to practice-based placements), thus supporting Maslow's claim that unless the lower-level human needs of security and belongingness are met, humans will be unable to develop their potential as effectively:

> It gave us a lot of confidence because we could constantly refer to each other and ask "Did I do that OK?".

> Every time you go through a crisis of confidence, it helps to have someone else who's suffering the same woes or at least understands them.

Some of the pre-service teachers in the pairings indicated that the collaborative relationship established in an early pairing was maintained throughout the duration of the initial teacher education programme, even when in subsequent solo practice-based placements. For these students, the importance of belongingness provided a fundamental basis upon which to develop as teachers. We have anecdotal evidence of these supportive collaborative relationships extending into the early years of teaching.

In comparison with solo placements, other benefits of paired placements identified by pre-service teachers, at Maslow's level one, included being able to adapt to a new environment more easily, and quickly establishing professional collaborative relationships with experienced teachers, both within the mathematics department and in the wider school. One pre-service teacher indicated that the paired placement

> helped with building relationships within the [mathematics] department. It made the whole thing very smooth and comfortable.

Organisation, record-keeping and necessary paperwork for the provider institution were also acknowledged, by both pre-service teachers and mathematics mentors, as being more efficient because of the pairing of students:

> There was always someone else there to remind you of things, as well, such as when the review statements were due in.

Working closely together in collaboration often required pre-service teachers to 'think on their feet', thus developing higher-level teaching skills, representing needs at Maslow's level two.

Balance of paired and solo teaching

Our partnerships took the view that, for students to derive the most benefit from a pairing, they should feel able to operate as *critical friends* in the collaborative relationship, confident to trust and confide in each other in a positive and receptive manner. Such a relationship reflects the interdependence we described earlier in this chapter but, in this case, between peer pre-service teachers. Pre-service teachers' reflections on their experiences of paired practice-based placements suggest that they developed a mutually respectful relationship, which supported their security needs at Maslow's level one. One student reported her experience of working in a shared lead situation in the classroom:

> Sometimes in the classroom, you think "Oh, what am I going to say now?" and the other one [pre-service teacher] steps in and automatically takes over.

It is in the role of *critical friends* that Maslow's level two needs are most evident. Pre-service teachers described situations in which their developing knowledge of each other's strengths and weaknesses enabled each to understand better what they needed to do in order to improve their own skills as a teacher. One example of this was when two pre-service teachers in a paired placement were interviewed for a teaching post in the practice placement to which one of them was appointed. Reflecting on the experience, the pre-service teacher not appointed said the paired placement had helped him to more clearly understand his peer's suitability for the particular post because he knew the strengths of his peer and recognised his own different but equal strengths. These processes impinge on Maslow's level three needs because these pre-service teachers are beginning to reflect on and interpret knowledge and understanding in order to know themselves better.

Being able to mutually plan lessons and closely observe each other in the implementation of these plans in the classroom allowed pre-service teachers to recognise and categorise competency-based progress towards qualified teacher status. Maslow's level two needs were further demonstrated in their accounts of how their exchanges of information about classes reflected what each of the pre-service teachers needed, rather than what experienced teachers offered them. Here, we reiterate what we asserted earlier, that pre-service teachers can feel vulnerable on practice-based placements, and this type of communication supports their confidence in being able to access the necessary information at appropriate times in their development of teaching skills. Alongside the idea of pre-service teachers working as *critical friends*, our practice-based partnerships also reflected a strong view that these novice teachers should have a distinctive element of *solo* teaching, in preparation for the solo teaching that will be expected of them in their first year of teaching. Thus, our practice-based partnership guidance recommends that every pre-service teacher should have sole responsibility for at least one class in every practice-based placement and offers several different models of allocation of classes to paired pre-service teachers.

One of these models is a development of Smith's (2004) tandem bicycle model, a lead and assistant model which emphasises the importance of both partners taking active and collaborative responsibility for the lesson. Important elements of all paired work are identified as:

- shared responsibility for planning for the lesson and for groups and individuals within this;
- clearly defined roles for each of the paired partners during the process of the lesson;
- collaborative planning as part of the planning and evaluation cycle (including roles and responsibilities in feeding back);
- development of assessment approaches (the lead teacher has a major role in assessment of learning);
- mutually agreed peer-evaluation mechanisms.

In support of Sorensen's (2004) claims, our evidence from evaluations and interviews also suggests that collaborative practice-based placements offer substantial benefits for students, pre-service teachers and practice-based partners. Pre-service teachers and mathematics mentors indicated that a pre-service pairing provides students with more individual attention; there is more expertise in the classroom for students to draw upon, and classroom management is improved. Pre-service teachers benefit from mutual support, both emotional and professionally developmental. Classroom observations of pre-service teachers indicate that risk-taking is increased, both in the paired situation and solo practice-based placement, if this is subsequent to the pairing. Pre-service teachers reported that the pairing offered them the security to step out of the 'known' and explore situations that they might not have explored on their own. We argue that this allows them to engage with needs at Maslow's level three, thus enabling them to access knowledge about themselves that is necessary for greater developmental progress as teachers. In both the formal evaluation evidence and interview evidence, pre-service teachers supported this claim, indicating that the emphasis on the collaborative practice and the degree of reflection required of them had benefits for their individual professional development. In situations where the pairing worked very well, this professional development was most evident:

> Sharing the lesson planning was really good. We had the option to team teach or teach alternate lessons, so we decided to team teach. I couldn't get over how good our styles matched – we complemented each other.

Shared lesson planning, in particular, was reported as providing pre-service teachers with the opportunity for enriched planning for students' learning. Even in a context where the relationship between two middle-aged male pre-service teachers was never fully established and culminated in parallel solo practice-based placements in the same mathematics department, there was room for effective reflection. One of these pre-service teachers reported:

> I think because we were both older people, we were both set in our ways and not quite so adaptable. I think I brought a lot of baggage into school and maybe this [the paired placement] helped me to see myself in a mirror. I learnt a lot about myself.

Both mathematics mentors and senior personnel in the practice-based partner sites reported benefits to the mathematics department and the school as a whole. Nokes and his colleagues (2008) reported similar benefits in a parallel study in the United States. They found that pre-service teachers developed a rich curriculum for their students by planning together. Increased risk taking, as an outcome of this enhanced planning and the security of being able to rely on another in the teaching context, is further evidence of how collaboration can increase teachers' willingness to engage in changing their teaching approaches, as suggested by Meirink and her colleagues (2010). Perhaps the more important evidence from the study by Nokes and his colleagues is that students reported "getting more help, appreciating hearing two points of view, receiving higher test scores, and enjoying activities more" (2008: 2174). Mentors indicated several positive outcomes of pairing pre-service teachers. The main factor was the impact on the quality of learning for their students. As indicated earlier, the adage that 'two heads are better than one' usually provided enhanced learning for students. Katharine Burn (1994), writing from the perspective of a mentor, claims:

> one of the indirect benefits of collaborative teaching is the opportunities presented by having two teachers in the classroom. It can allow for a much more lively or challenging presentation of ideas . . . It can make the management of learning activities much smoother . . . It can increase the amount of individual attention and support that pupils [students] receive, and can facilitate assessment or setting of differentiated tasks
>
> (1994: 6)

Mentors also indicated that mentoring two pre-service teachers did not require twice the time, so was more efficient, as the pre-service teachers relied on each other for organisation, record-keeping and other 'low-level' elements of their training. Considering the motives for engaging in practice-based training of pre-service teachers listed earlier, one of the benefits identified by mentors was the opportunity to 'cast an eye' over two potential teachers for the mathematics department.

Senior personnel stated that additional adult support in the practice-based institution meant that some external activities, requiring high levels of adult–student ratios, were ensured. What is critical in the reports of benefits to mathematics departments and the practice-based context was the development of stronger collaborative working relationships amongst experienced teachers in mathematics departments, increased sharing of resources, and more sharing of information about teaching styles between experienced teachers.

Final remarks

We have highlighted the importance and influence of mutual interdependence of collaborative ways of working in the mentor–mentee relationship, whether this is between two or between more than two adults. Such collaborative relationships also strengthen the assessment process for measuring performance of pre-service teachers against competence-based criteria. Close, interdependent collaboration enables the mentor not only to recognise the understandings, competences and professional skills of the pre-service teachers they mentor, but also to bring trustworthy 'insider' information to an assessment of these criteria for qualified teacher status. Proactive collaborative mentoring practices, based on reflective practice and informed by theoretical perspectives and ***craft knowledge***, can lead to transformative change for both the pre-service teachers and their mentors. Meirink and her colleagues (2010) argue that identifying a shared problem and developing a shared language to solve the problem facilitates this collaboration. Indeed, King (2006: 372) argues that, "Knowing about effective learning is more than simply observing classes, 'having a go' at teaching and then reflecting on outcomes. It includes extensive dialogue, talking with and listening to colleagues." The mentor has an important role in these conversations, particularly for pre-service teachers in paired placements, in maintaining clarity of purpose and in aligning developing knowledge with *craft knowledge*. In a piece of reflective writing, one of our pre-service teachers wrote:

> My initial impression of teaching was that it was quite a lonely profession in which the teacher had to work alone to teach his or her class. It [the research] has shown me that teaching is a collaborative process . . . Collaborating with fellow professionals is something I will rely on in the future. I was able to develop a support network within my school and this had a very positive impact on my professional development.

Through enabling collaboration with a range of teachers in the school, the practice-based mentor allowed this pre-service teacher to alter fundamental preconceived beliefs and attitudes about the nature of teaching mathematics.

References

Aderibigbe, S. (2011) 'Exploring collaborative mentoring relationships between teachers and student teachers', *Teacher Education Advancement Network Journal*, 3(1). Online. Available HTTP: <http://bit.ly/xMlqKB> (accessed 27 December 2012).

Burn, K. (1994) *Mentoring in Primary Schools. Collaborative teaching: a mentor's perspective, Reading 6*, Milton Keynes: Open University Press.

Hagger, H. & McIntyre, D. (1994) *Mentoring in Primary Schools. Collaborative teaching, Reading 7*, Milton Keynes: Open University Press.

King, S. (2004) 'Emerging models of teacher training in England, *International Research in Geographical and Environmental Education*, 13(2), 197–204.

King, S. (2006) 'Promoting paired placements in initial teacher education', *International Research in Geographical and Environmental Education*, 15(4), 370–86.

Lopez-Real, F. & Kwan, T. (2005) 'Mentors' perceptions of their own professional development during mentoring', *Journal of Education for Teaching*, 31(1), 15–24.

Maslow, A. (1970) *Motivation and Personality*, 3rd edn, Harlow: Essex Publishers.

Meirink, J., Imants, J., Meijer, P., & Verloop, N. (2010) 'Teacher learning and collaboration in innovative teams', *Cambridge Journal of Education*, 40(2), 161–81.

Nokes, J., Bullough, R., Egan, W., Birrell, J., & Merrell Hanson, J. (2008) 'The paired placement of student teachers: An alternative to traditional placements in secondary schools', *Teacher and Teacher Education*, 24(8), 2168–77.

Office for Standards in Education (2003) *Quality and Standards in Initial Teacher Training: Inspected 1999/2002*, HMI publication 546, London: Ofsted.

Schön, D. (1983) *The Reflective Practitioner*, New York: Basic Books.

Smith, J. (2004) 'Developing paired teaching placements', *Educational Action Research*, 12(1), 99–125.

Sorensen, P. (2004) 'Learning to teach collaboratively: the use of subject pairs in the school practicum', *Canadian Journal of Educational Administration and Policy*, 32.

Sorensen, P., Hoult, L., Philpott, C., & Katene, W. (2002) 'Peer learning in Initial Teacher Education (ITE): The use of subject pairing in school experience placements as a strategy for professional learning' Paper presented to the *Annual Conference of the British Educational Research Association*, University of Exeter, England.

Training and Development Agency (2008) *Standards for the Award of Qualified Teacher Status*, London: TDA.

Ventegodt, S., Merrick, J., & Andersen, N. (2003) 'Quality of Life Theory III: Maslow revisited', *The Scientific World*, 3, 1050–7.

Wallis, T. (2012) 'Building partnerships for success: a model for working with teacher trainers, mentors and trainees', *Teacher Education Advancement Network Journal*, 4(1). Online. Available at HTTP: <http://bit.ly/I5VJ5s> (accessed 27 December 2012).

Wenger, E. (1998) *Communities of Practice: Learning, Meaning and Identity*, Cambridge: Cambridge University Press.

Wilson, P. & Bolster, A. (2011) *New Models of Teacher Education: Collaborative paired placements*, Bristol: ESCalate.

Wilson, P. & Edwards, J. (2009) 'Paired ITE teaching placements: implications for partnership development'. *Proceedings of the British Society for Research into Learning Mathematics*, 29(2), 82–7.

Chapter 7

Developing the potential of all pre-service mathematics teachers

Patrick Barmby and Ruth Edwards

This chapter addresses the meaning of 'potential' with regard to pre-service mathematics teachers and the identification and development of this potential by practice-based mentors. Consideration is given to the early identification of strengths and weakness in areas that are key attributes for teachers, such as teacher knowledge, and describing how practice-based mentors can provide experiences to support development in these areas. We discuss the ways in which both the provider institution and practice-based mentors can work together across the range of teaching, tasks and activities assessed, including considering how best to give appropriate feedback to pre-service teachers. Use of feedback to actively encourage reflective practice is addressed and the need for practice-based mentors to use a diversity of approaches.

Introduction

We begin with the premise that, prior to beginning their initial teacher education programme, pre-service teachers and those who select them for such programmes have identified some sort of 'potential' for teaching. For many prospective teachers, some kind of 'teaching type' activity, coaching, mentoring, tutoring or other way of working with young people or their peers is a key factor in their career choice. For those intending to teach at secondary level, there will also be some rationale for identifying mathematics as the subject they have the potential to teach, although this is likely to be less true for many of those wanting to teach at primary level. This chapter, therefore, addresses the meaning of 'potential' with regard to pre-service mathematics teachers and the identification and development of this potential by practice-based mentors. The development of pre-service teachers during the period of their initial teacher education through to their early career following qualifying is also explored.

Context

The need for high-quality mathematics teachers has been an on-going issue for over thirty years. In 1982, the Cockcroft report identified that "no efforts to improve the quality of mathematics teaching are likely to succeed unless there is an adequate supply of suitably qualified mathematics teachers" (1982: 806). Two decades later, Sir Gareth Roberts noted in his report a number of deep-seated issues common to schools and further education, one of which was "shortages in the supply of physical science and mathematics teachers/lecturers" (Roberts 2002: 3). In 2010, the Department of Education for England reported that 16 per cent of mathematics lessons in secondary schools were not taught by a specialist (DfE 2010: 7) and data for later that year showed that 19 per cent of all secondary teaching vacancies were for teachers of mathematics (DfE 2011a). In the Czech Republic more than 15 per cent of secondary mathematics pre-service teachers interviewed as part of an international comparative study of mathematics initial teacher training stated that they only intended to teach for the first year following their graduation and a further 20 per cent had decided before graduation that they would never teach. Similar figures were reported in the Ukraine, with 15 per cent of pre-service mathematics teachers reporting that they did not intend to teach at all (Burghes 2011a: 12–13). The National Center for Education Statistics (2011) projects a need for an increase in the number of teachers employed in schools in the United States, and in some Australian states between 25 per cent and 40 percent of beginning teachers are likely to leave the teaching profession in the first five years (Ewing & Smith 2003).

In response to concerns about recruitment and retention of teachers, one might suggest that 'potential to teach' is not being accurately identified, or that entry requirements have not been set at the correct level (DfE 2011b). It might also be that the identified 'potential' is not being realised during training. Barber and Mourshed (2007) analysed twenty-five of the world's school systems, ten of which were the best performing as defined by the OECD's Programme for International Student Assessment (PISA). They subsequently identified three key factors which had the greatest impact on the performance of students in mathematics:

- getting the right people to become teachers;
- developing them into effective teachers;
- ensuring that the system is able to deliver the best possible teaching for every child.

Early identification of strengths and weaknesses in areas that are key attributes for teachers, for example teacher knowledge, supports the first of these important factors. In addition, effective mentoring where practice-based mentors provide experiences to support development will support the second of these factors. Both of these contribute to reducing the rate of attrition of mathematics teachers from the profession. We therefore start by exploring the first of these factors, relating the identification of the 'right people' to the notion of 'potential' of teachers. We then examine how we can further build upon this potential by developing pre-service teachers into effective practitioners through the mentoring provided to them in their training.

Identifying 'the right people'

What does it mean to 'get the right people to become teachers'? Barber and Mourshed (2007) identified three characteristics relating to individual teachers which improve the quality of teaching in classrooms. Teachers need to

- gain an understanding of best practices;
- become aware of specific weaknesses in their practice;
- be motivated to make changes.

Developing teachers' potential is to develop their ability to work towards and achieve the above characteristics. The characteristics that are needed by the 'right people' are not only having the knowledge of best practices (and we note here that this is knowledge of practice rather than knowledge of subject), but also the ability and motivation to identify and act upon weaknesses in this practice. Therefore, we can begin a more in-depth, step-by-step exploration of how we can develop the potential of pre-service teachers, specifically in the subject of mathematics, based around the three required characteristics of teachers identified.

The role of teacher knowledge and practice in defining potential

Beginning with the development of teachers' understanding, we first consider the nature of this understanding. In his influential paper, Shulman (1986) strongly emphasised the understanding required by teachers: "With Aristotle we declare that the ultimate test of understanding rests with the ability to transform one's knowledge into teaching. Those who can, do. Those who understand, teach" (1986: 14).

Shulman clarified the nature of this knowledge to be transformed, putting forward three types of 'content' knowledge required within any given subject: subject matter content knowledge, pedagogical content knowledge and curricular knowledge. As indicated in Chapter 1, subject matter content knowledge includes the factual content of the subject and how the subject functions as a discipline (for example, the notion of proof in mathematics). Pedagogical content knowledge is the knowledge required for specifically teaching the subject, being distinct from general pedagogical knowledge (Shulman 1987). Shulman (1986) proposed the following components to pedagogical content knowledge: how we represent information, including the most powerful representations (including models, illustrations, analogies and examples) for teaching, and also an understanding of how students generally learn the subject (including what makes topics easy or hard, typical conceptions and misconceptions, and how to tackle typical misconceptions). Finally, curricular knowledge consists of knowledge of teaching programs and use of teaching materials.

In clarifying the nature of this understanding for teachers, we can see that the emphasis is not just on subject knowledge *per se* within mathematics, but also on the broader knowledge related to practice and the pedagogical knowledge required to teach mathematics. Research has also highlighted the importance of the 'metacognitional' knowledge required by teachers that enables them to reflect on their practice. This moves us to the second of the characteristics identified by Barber and Mourshed (2007). Returning to Shulman (1987) once again, he additionally highlighted the process by which teachers reflect upon and develop their knowledge and understanding of teaching mathematics. They may begin by *comprehending* the mathematical knowledge to be taught; this knowledge is then *transformed* by the teachers into a form of *instruction* for the students. The success of this instruction can be *evaluated*, and in light of this, the teacher can *reflect* on how to develop their knowledge and understanding of mathematics and teaching mathematics. Such meta-reflection is explored in Chapter 12. More detail on this process of reflection is also examined by Pollard (1997), who identified a cyclical process whereby teachers evaluate and revise their practices continuously. He also identified six key characteristics of reflective practice in which this:

- requires an active concern with aims and consequences;
- requires an attitude of open-mindedness;
- requires responsibility and wholeheartedness;
- requires competence in methods of classroom enquiry;
- is based on teacher judgement informed by self-reflection and also the educational discipline; and,
- is enhanced through collaboration and dialogue with colleagues.

The emphasis on self-reflection alongside the education discipline and the collaboration with colleagues mirrors Shulman's (1987) 'knowledge bases' for developing the understanding of teachers, where personal experiences, educational materials and advice (including that of colleagues) and scholarly research can be drawn upon in order to reflect upon and inform practice.

In order for this active reflection to take place, as was highlighted in the third characteristic by Barber and Mourshed (2007), the teacher needs to have the motivation to carry out this process of reflection and enquiry. Crawford and Adler support this, saying:

> Only through active engagement with problems and questions that are personally meaningful to them will they develop a rationale for action. Only through understanding their own learning through research, inquiry, investigation, and analysis will they come to understand such processes among students in their care.
>
> (1996: 1201)

We have now identified the characteristics that we need to focus upon in this process, and therefore what we need to be focusing on in initial teacher education. We have identified the importance of the development of teacher knowledge in the broadest sense, encompassing the knowledge of subjects and also knowledge of how this is transformed into instruction, through encouraging teachers' abilities to reflect on this process. We have also identified how this process of reflection can be supported, not just through reflection on personal experiences in the classroom, but enhanced through dialogue and advice from colleagues, using research to inform these reflections, and methods of classroom enquiry to identify and examine more deeply the issues at hand. This entire process can only be achieved if teachers are engaged and motivated to undertake these tasks.

Identifying potential

Academic performance is one factor that needs to be considered when recruiting pre-service teachers. One might think that increasing the academic entry requirements for initial teacher education programmes is one way of more accurately identifying those with 'potential' to teach. However, the support from educational research generally for this emphasis on prior academic or subject knowledge qualifications is somewhat lacking. Despite the common use of 'scholastic aptitude' measures in countries such as the United States, the relationship between such measures and success in teaching is not strong (Schalock 1979). In fact, Freeman, Martin, Brousseau and West (1989) found that those about to enter teacher training with higher or lower academic qualifications differed little in terms of beliefs about teaching and commitment to the profession. More specifically in mathematics, research over time has highlighted the apparent lack of impact that previous qualifications in mathematics have had on the teaching of the subject (Eisenberg 1977; Ball 1990; Askew 1999; Kahan, Cooper & Bethea 2003).

Therefore, there seems little support for this focus on prior academic qualifications in order to identify prospective teachers. Furthermore, one of the possible problems faced by the initial teacher education programmes in terms of their recruitment processes is the lack of connection between programme intentions or outcomes and the selection criteria being used (Applegate 1987). If our intended outcome for recruitment to initial teacher education programmes is to develop the potential of pre-service teachers, then our selection processes need to be more in line with this intention. Therefore, in selecting potential teachers for our programmes, we need to focus on these broader attributes of applicants.

Focusing on pre-service teachers' motivations and ability to reflect on practice, then, Giovannelli (2003) demonstrated that a reflective disposition towards teaching is indeed related to effective teaching. Taking this further in terms of teacher recruitment, one of the authors of this chapter examined the results of an interview-based teacher selection process, looking specifically at candidates' abilities to reflect on their experiences in schools so far (Barmby, Bolden, Higgins & Tymms 2012). This was carried out for those applying to an undergraduate degree in Primary Education, and compared the outcomes for pre-service teachers (both degree classification attained and also final grading for school practice) against Advanced level grades from their own schooling, and also scores obtained from the admission interviews. These interviews consisted of a five-minute presentation, followed by three specified questions exploring the students' reflections on their experiences in primary schools (all candidates had been asked to arrange two weeks' experience in schools prior to interview). The students were graded 1 = excellent, 2 = good, 3 = satisfactory or 4 = reject for their interviews by two interviewers after discussion between them. The study found that, as might be expected, Advanced level grades correlated significantly with final degree classifications for these pre-service teachers. Advanced level grades, on the other hand, did not correlate significantly with the final grading from school practice. This is in line with the findings from the research given above with regard to academic measures of aptitude. The interview grades, however, were found to correlate significantly with both final degree classification and final gradings from school practice. Therefore, the measure related to pre-service teachers' ability to reflect on their school experiences was a better predictor of effective teaching as measured by these final gradings.

What we are suggesting here is that the consideration of pre-service teachers' potential, in terms of their ability and motivation to reflect and develop their teaching and their understanding of practice, is important not only during initial teacher education programmes, but also in terms of selecting potential teachers. Perhaps further future consideration of this idea of 'potential' will help tutors and mentors to tackle the problematic issue of teacher recruitment criteria.

Developing pre-service teachers' potential through teacher education

Having associated the notion of 'potential' with the ability and motivation to reflect upon and develop one's teaching, we can look further at how to further develop this potential in pre-service teachers during their initial teacher education programme. What the research literature identifies is that this ability to reflect will not develop in isolation. Calderhead (1989) highlighted that pre-service teachers need to develop 'metacognitive analytic skills' in order to reflect on their experiences, as highlighted in Chapter 12. Likewise, Yost, Sentner and Forlenza-Bailey (2000) stated that pre-service teachers lacking guidance in this area could not apply their pedagogical learning in ways to enhance their teaching. Hatton and Smith (1995) highlighted the time required to develop these essential metacognitive skills. In addition to developing pre-service teachers' skills in reflecting on their practice, as identified previously, this goes hand in hand with developing the motivation to reflect. It may be that identifying one's own weaknesses through reflection may lead to feelings of vulnerability (Hatton & Smith 1995). Cole (1997) identified other emotional 'impediments' to reflection, such as anxiety, fear and helplessness. In developing both the skills and the motivation to reflect, the research identifies the importance of collaborative approaches, or the use of the 'critical friend', to support these processes: "A powerful strategy for fostering reflective action is to engage with another person in a way which encourages talking with, questioning, even confronting, the trusted other, in order to examine planning for teaching, implementation, and its evaluation" (Hatton & Smith 1995: 41).

Yost, Sentner and Forlenza-Bailey (2000) also identified discourse methods, as well as constructivist action research and writing methods, for encouraging reflection. Similarly, Malm (2009: 86), citing Pickering, Daly and Pachler (2007), highlighted shared practice, collaborative networks and scholarly reflection on practice as being useful tools. In the following section, we develop these themes of skills and motivation to reflect in initial teacher education programmes, looking particularly at the role of the mentor in this process.

The role of the mentor in developing potential

Pre-service teachers have contact with a number of individuals throughout the duration of their course: supervising tutors, school mentors, departmental and school colleagues, other pre-service teachers on their course, the learners they teach and significant others at home; the influence of each of these will vary. Hobson *et al.* (2008) identified that it was important to pre-service teachers that they 'get on' with the people they are training and working with and that the extent to which they perceive that their mentor has time for them has a significant impact on their achievement and success whilst on school placement. One of the most frequently cited reasons pre-service teachers give for leaving their initial teacher education programme is lack of support from their school mentor (Chambers, Coles & Roper 2001). Research has shown (Hobson 2002) that pre-service teachers highly value a timetabled weekly meeting with their school mentor and having a mentor who is prepared and able to make time to discuss matters of mutual interest and concern. Qualities also valued in mentors were assisting them (the pre-service teachers) to develop the ability to manage learners and maintain discipline, to plan lessons, to use a range of teaching methods effectively, to develop their subject knowledge, in addition to observing lessons and in providing feedback.

Learning to become a teacher covers many aspects of both theory and practice and it can often be confusing for a pre-service teacher to identify how these fit together. In some initial teacher education programmes pre-service teachers are required to produce written assignments for an awarding institution. Pre-service teachers sometimes perceive these assignments as the least useful part of their training (Hobson *et al.* 2008). As mentioned earlier, the mentor is not the only person who has an impact on the success of a pre-service teacher; however, they have been found to be the most significant (Chambers *et al.* 2001). A mentor with a clear picture of the structure of the training and an overview of what is covered in the more theoretical parts of the programme can support a pre-service teacher by helping them to link theory with practice:

- attend meetings arranged between supervising tutors and mentors so that there is good communication between those with different roles within the

programme and everyone understands the contributions of each part of the programme;
- support learning by linking targets to the more theoretical parts of the programme and giving opportunities to put theory into practice;
- become involved with delivering some of the more theoretical parts of the programme. Pre-service teachers rate highly those parts of their taught programme which are delivered by school mentors.

The value of school placement experience in initial teacher education is self-evident, as is the value of the support of a teacher mentor for the pre-service teacher. This support can be critical to the success of the pre-service teacher as relationships make important contributions to the induction experiences of beginning teachers (Carter & Francis 2001). In a review of the international research literature on mentoring beginning teachers, Hobson *et al.* (2009) identified a number of benefits perceived by both mentees (pre-service teachers) and mentors. For the pre-service teacher these included the provision of emotional and psychological support, guided development of teaching capabilities and socialisation into the profession. Mentors reported positive personal benefits which contributed to their own professional and career development. These aspects are further developed in Chapters 2 and 6. These included learning new skills from the pre-service teachers they were mentoring, improved personal self-reflection, increased confidence in their own teaching and pride in seeing their mentee succeed. Conversely, some mentors reported that they felt threatened by pre-service teachers observing their lessons, isolated in the role and restricted by lack of time to manage their workloads. These findings in turn are reflected in the disadvantages noted by the pre-service teachers: lack of emotional support, feeling insufficiently challenged by their mentors, having little attention paid to pedagogical issues or explicit links made between theory and practice. Such practice can lead to pre-service teachers either leaving their initial teacher education course or simply 'fitting in' and conforming with traditional teaching methods. One of the purposes of this book is to explore ways in which mentors and their mentees can construct positive experiences for both parties to address these issues.

Developing reflection

As well as the general role that mentors play in supporting pre-service teachers, we can examine their specific role in terms of developing the reflective skills of pre-service teacher, and also their motivation to reflect. Related to this first issue, Burghes's (2011b: 15) study suggests that areas of concern for both primary and secondary pre-service mathematics teachers are: challenging able learners and supporting less-able learners, planning, management of resources, class management, use of IT, teaching strategies, pedagogy and mathematical knowledge. As experienced teachers are aware, managing these concerns works in tandem with effective lesson planning, evaluation of and reflection on these plans. Lesson planning is an extremely complex activity and whilst a template can be given to a pre-service teacher to follow, it is the minutiae of detail, understanding of the topic and the individuals to whom the lesson is to be taught that lead to a successful lesson. Pre-service teachers undoubtedly need much support in this area, some strategies for which are set out below:

- a group of pre-service teachers are given a topic whereby they plan a lesson together. They can subsequently either all teach the same lesson or re-write it in their own style and teach their individual version. If this task is undertaken early in the school placement it gives a good tool for discussion and enables a mentor to get a better 'feel' for the strengths of the pre-service teacher;
- the pre-service teacher plans a lesson for the mentor to teach and then observes the lesson;
- the mentor jointly plans a lesson with the pre-service teacher and then one or both of them teaches it whilst the other observes, leading to a discussion of any differences in outcome.

Each of these suggestions has within it an element of 'lesson study', an increasingly used tool for professional development. Originating in Japan, 'lesson study' involves a group of teachers who jointly plan a lesson which one of them subsequently teaches. The others in the group observe the lesson, make notes and then the group meets to discuss and reflect on the lesson. Often the teachers involved will re-write the lesson following their discussions and another of the group will teach the amended version. If you are fortunate enough to have more than one pre-service teacher working with you, you may like to try following a pattern of lesson study with them where you plan, teach, observe, reflect, critique and re-plan together.

Another aspect of the mentor's role that supports reflection is the observation of lessons and the subsequent feedback to the pre-service teacher. Feedback can be informal such as a chat at the end of the lesson, rough written notes, a formal lesson observation or form part of the agenda for the weekly mentor meeting. Some points for consideration are:

- for informal written lesson observations the pre-service teacher may have a notebook that all teachers observing their lessons can write in. In this way, observers can see what others have written and common areas for development may become apparent;
- on written feedback, highlight or write in bold the key points;
- make constructive criticisms because identifying what is good will boost the self-confidence of the pre-service teacher. Refer to positive points first, for example, by praising the worksheets a pre-service teacher has spent time preparing but also giving suggestions and friendly challenges to improve and develop them. Rather than asking "how did you think the lesson went?" at the end of a lesson observation, open with "What went well in the lesson – how could you make it even better?" "What would you do differently?"

- set targets at the end of a lesson observation and set a time to follow these up to see how well they are being met. Identify achievable targets and share them with the other teachers in the department;
- offer clear and practical advice that relates to specificity of action.

These observations and feedback can be used to tackle specific issues faced by the pre-service teacher. A survey of pre-service secondary mathematics teachers at the University of Southampton in 2010 identified three main areas of difficulty common to those who were struggling on their initial teacher education programme: isolation on school placement, difficulty developing a presence in the classroom, and struggling with lesson planning and lack of access to resources (Edwards & Hyde 2011). A mentor can provide support for pre-service teachers needing to develop their classroom presence by outlining to the pre-service teacher a number of strategies that they can incorporate into their reflections on their teaching such as:

- considering how much of the classroom they can 'see';
- making sure they 'own' the classroom; that they move around the classroom and don't just stay at the front;
- thinking about varying the tone/pitch/speed of speech. The pre-service teacher and mentor could try going to a sports hall together to try out different levels of speech, or visit other classroom arrangements such as a computer suite or laboratory to see how the room affects their voice;
- learning the names of their students;
- having one-to-one conversations with learners;
- reprimanding in a positive tone.

To an experienced teacher many of these suggestions may appear to be obvious; however, it is sensible to bear in mind that the style and degree of mentoring provided and the accompanying reflection needs to be modified according to a pre-service teacher's different stages of development. When studying initial teacher education programmes featuring two school placements, Hobson (2002) found pre-service teachers generally rated their first mentor as more effective than their second. He identified this as possibly being due to mentors providing more assistance to pre-service teachers on their first placement. An alternative explanation might suggest that mentors had not modified their ways of working to match the differing needs of pre-service teachers in their second placement, and so were seen to be less effective. Therefore, the areas in which, and the methods by which, mentors support pre-service teachers' reflections need to vary according to their stage of training.

The affective dimension of reflection

In addition to developing the process of reflection, mentors need to support pre-service teachers' motivations to reflect. As teachers, we have all experienced those feelings of satisfaction when a learner says "oh I get it now" or "thanks for a great lesson" when they leave the classroom. Conversely, there have been times when we have groaned about writing reports, begrudged staying late for meetings or felt let down by lack of support from a colleague. This range of positive and negative emotions is mirrored by pre-service teachers in respect of their relationships with learners, mentors and other teachers in their placement school, their work–life balance and their own sense of development and efficacy as a teacher (Hobson *et al.* 2008). Being aware of the mix of highly aroused emotions that pre-service teachers encounter in the course of their initial teacher education, and understanding how to recognise and respond to them, is an important part of the role of the mentor.

In our experience, there are identifiable points in our initial teacher education programmes where pre-service teachers commonly experience additional stress and feel that they are not coping well. One such point occurs for our pre-service teachers shortly after they have changed school placements and have also increased the number of lessons they are teaching each week. Our understanding is that they (mainly) have positive memories of the lessons they taught during their first school placement: they knew the students' names, which ones they needed to pay extra attention to and who in the year team they could go to for help; they 'knew how to teach'. However, all of a sudden they find they have less time, they don't know all the students' names or who they need to keep a special eye on. They don't like to ask for help because they feel they should know what they are doing by now – after all, they have already taught and had responsibility for their own class(es) previously. They panic, spend too long looking for resources trying to make every lesson 'perfect' and not long enough on careful planning and reflection. Lessons don't go as well as expected and the downward spiral begins.

When emotional changes are likely to occur is dependent on the structure and nature of the programme as well as the nature of the pre-service teacher (see the work on zones of promoted action in Chapter 8). Emotional stability during difficult times is in turn related to the experiences encountered by the pre-service teacher during their first school placement. Hayes (2003) found that, depending in part on their earlier school experiences, primary pre-service teachers also underwent similar enabling or disabling emotions before and during their second school placement.

Strategies for support at these times of emotional turbulence will vary; often pre-service teachers are reluctant to admit directly to their mentors that they are undergoing such emotions. Edwards and Hyde (2011) described setting up a social networking space for their secondary mathematics pre-service teachers and posted a comment about this period of the programme, explaining that, at that time, uncertainty was a common feeling to be experiencing. The online conversation that followed indicated that, in fact, a number of the pre-service teachers were suffering from a lack of self-conviction and were subsequently comforted by the knowledge that they were not alone in that respect. Some mentors have used drawings to elucidate the emotions that their pre-service teachers are experiencing by asking them to draw an image of themselves with their classes at various stages in the course. These drawings subsequently form the basis of a dialogue between the mentor and the pre-service teacher, supporting the process of reflecting on the pre-service teachers' experiences. Swennen, Jorg and Korthagen (2004) found that drawings formed a useful tool for measuring the development of a teacher. For some pre-service teachers there was an intensity of emotion represented within their drawings that enabled them to remember in detail their thoughts and emotions at the time several months later. In addition to the strategies presented above, it is important that mentors are active listeners. Having time for, and listening to the concerns of a pre-service teacher, being supportive, reassuring and non-judgemental, offering advice and guidance, particularly at identifiable times of stress, is a key aspect of the mentor's role in emotionally supporting pre-service teachers in their reflections on their experiences.

Moving from developing potential to continuous professional development

For pre-service teachers, the learning journey to becoming a teacher does not stop at initial qualification. In many countries those beginning their teaching are supported by programmes of induction. Wong, Britten and Ganser (2005: 379) define induction as a

> highly organized and comprehensive form of staff development, involving many people and components, that typically continues as a sustained process for the first two to five years of a teacher's career. Mentoring is often a component of the induction process.

In its most basic form it is for the newly qualified teacher to consolidate what they have learned and demonstrated during their initial teacher education. However, there is much more that can be given to a newly qualified teacher through effective mentoring during this period. Mentoring is a long-term process with a focus on the support of the career development and growth of the individual. As such, the mentoring of newly qualified teachers is an on-going process throughout not only their induction year but beyond. Ewing and Smith (2003: 30) in their review of teacher education in New South Wales, Australia, recognised "the urgent need to provide well managed, effective induction programmes at school, district and system level that go beyond the first year of teaching and are based in the professional learning needs of beginning teachers". More recently Curtis (2012) cited the findings of a nationwide study of more than 1500 middle school and high school mathematics teachers in the United States which identified a significant positive relationship between mathematics teachers who had participated in an induction or mentoring program and the decision to remain in the teaching profession.

Burghes (2011a: 50) makes a number of recommendations for the support mathematics teachers should receive in their first year of teaching, with a particular focus once again on supporting teachers' reflections on their experiences:

> a significantly reduced timetable, enabling them to have time for lesson preparation, reflection, working with expert teachers (through a lesson study model) and with university staff on practical work contributing to modules in a Master's Degree; this should continue for at least three years, with a gradual increase in teaching time.

He also suggests that a positive outcome of this more suitable support in school is the increase in the numbers of newly qualified mathematics teachers remaining in teaching. Likewise, in their critical review of the research into the impact of induction and mentoring programmes for beginning teachers, Ingersoll and Strong (2011) found that participation in induction had a positive effect in many areas of newly qualified teachers' professional development. Higher satisfaction, commitment and retention, better performance at various aspects of teaching and higher student achievement were all noted. Programmes of induction vary between different educational systems, as does the role that the induction mentor plays within these programmes. Supportive mentors can help the beginning teacher by devoting their time, experience, understanding and enthusiasm for education, by developing an understanding of best practices by working with and guiding the newly qualified teacher, reflecting on practice and discussing areas for improvement and being a positive role model by demonstrating their own motivation to make change. Looking back at our earlier description of developing the potential of teachers, this final section has brought us in a full circle back to where we began in this chapter when we defined potential. We can see that this process of developing potential, in terms of reflecting and improving upon practice, and having the motivation to do so, is important not only at the beginning of pre-service teaching, but continues into the subsequent career of the newly qualified teacher.

References

Applegate, J. H. (1987) 'Theme: teacher candidate selection an overview', *Journal of Teacher Education*, 38(2), 2–6.

Askew, M. (1999) 'It ain't (just) what you do: Effective teachers of numeracy', in I. Thompson (ed.), *Issues in Teaching Numeracy in Primary Schools*, Buckingham: Open University Press, pp. 91–102.

Ball, D. L. (1990) 'Elementary and secondary teachers' understanding of division', *Journal for Research in Mathematics Education*, 21(2), 132–44.

Barber, M. & Mourshed, M. (2007) *How the World's Best-performing School Systems Come Out on Top*, London: McKinsey & Company.

Barmby, P., Bolden, D., Higgins, S., & Tymms, P. (2012) 'Predicting success on a

Bachelor of Arts primary education degree course', *Educational Research*, 54(3), 323–30.
Burghes, D. (2011a) *Research Paper: International Comparative Study in Mathematics Teacher Training: Enhancing the training of teachers of mathematics*, Reading: CfBT Education Trust.
Burghes, D. (2011b) *Research paper: International Comparative Study in Mathematics Teacher Training: Recommendations for initial teacher training in England*, Reading: CfBT Education Trust.
Calderhead, J. (1989) 'Reflective teaching and teacher education', *Teaching and Teacher Education*, 5(1), 43–51.
Carter, M. & Francis, R. (2001) 'Mentoring and beginning teachers' workplace learning', *Asia-Pacific Journal of Teacher Education*, 29(3), 249–62.
Chambers, G., Coles, J., & Roper, T. (2001) 'Students withdrawing from PGCE (secondary) courses'. Paper presented to the British Educational Research Association Annual Conference, University of Leeds, 13–15 September.
Cockcroft, W. H. (1982) *Mathematics Counts*, London: HMSO.
Cole, A. L. (1997) 'Impediments to reflective practice: toward a new agenda for research on teaching', *Teachers and Teaching: Theory and Practice*, 3(1), 7–27.
Crawford, K. & Adler, J. (1996) 'Teachers as researchers in mathematics education', in A. J. Bishop, K. Clements, C. Keitel, J. Kilpatrick & C. Larborde (eds), *International Handbook of Mathematics Education*, Dordrecht: Kluwer, pp. 1187–205.
Curtis, C. (2012) 'Recruiting and mentoring mathematics teachers', *International Journal of Humanities & Social Science*, 2(18), 39–44.
DfE (2010) *The Case for Change*. Online. Available HTTP: <https://www.education.gov.uk/publications/standard/AbouttheDepartment/Page1/DFE-00564-2010> (accessed 20 December 2012).
DfE (2011a) *Statistical First Release: School Workforce in England November 2010 (Provisional)*. Online. Available HTTP: <http://www.education.gov.uk/rsgateway/DB/SFR/s000997/index.shtml> (accessed 20 December 2012).
DfE (2011b) *Training the Next Generation of Teachers: an implementation strategy for discussion*. Online. Available HTTP: <https://www.education.gov.uk/publications/standard/publicationDetail/Page1/DFE-00054-2011> (accessed 20 December 2012).
Edwards, R. & Hyde, R. (2011) *Supporting STEM: Providing intervention strategies for borderline ITE students in STEM subjects to increase their success*, Escalate funded project. Online. Available HTTP: <http://escalate.ac.uk/8139> (accessed 20 December 2012).
Eisenberg, T. A. (1977) 'Begle revisited. Teacher knowledge and student achievement in algebra', *Journal for Research in Mathematics Education*, 8(3), 216–22.
Ewing, R. A. & Smith, D. L. (2003) 'Retaining quality beginning teachers in the profession', *English Teaching: Practice and Critique*, 2(1), 15–32.
Freeman, D. J., Martin, R. J., Brousseau, B. A., & West, B. B. (1989) 'Do higher program admission standards alter profiles of entering teacher candidates?', *Journal of Teacher Education*, 40(3), 33–41.
Giovannelli, M. (2003) 'Relationship between reflective disposition toward teaching and effective teaching', *The Journal of Educational Research*, 96(5), 293–309.
Hatton, N. & Smith, D. (1995) 'Reflection in teacher education: Towards definition and implementation'. *Teaching and Teacher Education*, 11(1), 33–49.
Hayes, D. (2003) 'Emotional preparation for teaching: A case study about trainee teachers in England', *Teacher Development*, 7(2), 153–71.

Hobson, A. J. (2002) 'Student teachers' perceptions of school based mentoring in Initial Teacher Training (ITT), mentoring and tutoring', *Partnership in Learning*, 10(1), 5–20.

Hobson, A. J., Ashby, P., Malderez, A., & Tomlinson, P. D. (2009) 'Mentoring beginning teachers: what we know and what we don't', *Teaching and Teacher Education*, 25(1), 207–16.

Hobson, A. J., Malderez, A., Tracey, L., Giannakaki, M., Pell, G., & Tomlinson, P. D. (2008) 'Student teachers' experiences of initial teacher preparation in England: core themes and variation', *Research Papers in Education*, 23(4), 407–33.

Ingersoll, R. M. & Strong, M. (2011) 'The impact of induction and mentoring programs for beginning teachers: A critical review of the research', *Review of Educational Research*, 81(2), 201–33.

Kahan, J. A., Cooper, D. A., & Bethea, K. A. (2003) 'The role of mathematics teachers' content knowledge in their teaching: A framework for research applied to a study of student teachers', *Journal of Mathematics Teacher Education*, 6, 223–52.

Malm, B. (2009) 'Towards a new professionalism: enhancing personal and professional development in teacher education', *Journal of Education for Teaching*, 35(1), 77–91.

National Center for Education Statistics (2011) *Projections of Education Statistics to 2020.* Online. Available HTTP: <http://nces.ed.gov/programs/projections/projections2020/sec3a.asp> (accessed 20 December 2012).

Pollard, A. (1997) *Reflective Teaching in the Primary School*, London: Cassell.

Roberts, G. (2002) *SET for Success. The supply of people with science, technology, engineering and mathematics skills (The Report of Sir Gareth Roberts's Review)*, London: HM Treasury.

Schalock, D. (1979) 'Research on teacher selection', *Review of Research in Education*, 7, 364–417.

Shulman, L. S. (1986) 'Those who understand: Knowledge growth in teaching', *Educational Researcher*, 15(2), 4–14.

Shulman, L. S. (1987) 'Knowledge and teaching: Foundations of the new reform', *Harvard Educational Review*, 57(1), 1–23.

Swennen, A., Jorg, T., & Korthagen, F. (2004) 'Studying student teachers' concerns, combining image based and more traditional research techniques', *European Journal of Teacher Education*, 27(3), 265–83.

Wong, H. K., Britton, T., & Ganser, T. (2005) 'What the world can teach us about new teacher induction', *Phi Delta Kappan*, 86(5), 379–84.

Yost, D. S., Sentner, S. M., & Forlenza-Bailey, A. (2000) 'An examination of the construct of critical reflection: Implications for teacher education programming in the 21st century', *Journal of Teacher Education*, 51(1), 39–49.

Chapter 8

Developing the potential of outstanding beginning teachers

Julie-Ann Edwards and Danielle Townsend

This chapter argues the case for developing the very best in the highest-performing pre-service and beginning mathematics teachers, firstly as an entitlement for the teacher concerned, but also to ensure an entitlement for all the students these outstanding beginning teachers will teach in the future. The role of *critical moments* as triggers in developing outstanding pre-service mathematics teachers is examined. Using a model of stages of expertise in teaching, and Vygotskian socio-constructivist approaches to learning through the *zone of proximal development*, the chapter supports practice-based mentors in identifying appropriate challenges and in encouraging a culture of risk-taking, in order to learn, as an expectation of these outstanding beginning mathematics teachers.

Introduction

As in the classroom, where students show an aptitude for mathematics, pre-service or beginning teachers can also demonstrate an aptitude for teaching mathematics, what is commonly termed a 'natural talent for teaching'. Such novice teachers pose different, but equally demanding, challenges for the mentor on both pragmatic and emotional levels. Orland-Barak and Yinon (2005) describe how the expertise of experienced mentors is highly influenced by the personal interactions they encounter with their mentees and the nature of the mentoring situation within which they find themselves. Mentors can, thus, experience situations where they act in ways ranging from a novice mentor to an expert advisor. Working in a mentoring capacity with a talented pre-service teacher is likely to increase the spectrum of Orland-Barak's and Yinon's description of regressions and progressions experienced in the mentoring role. We argue that recognising different types of 'talent' and working actively with this 'talent', as a mentor, can reduce the amount of regression experienced, so making the mentor–mentee relationship more stable and positive.

Naturally talented novice teachers can potentially pose some of the most difficult situations for the mentor. First, an assessment of strengths and consequent targeting of areas for development is more problematic when novice teachers appear to cope well in a variety of teaching contexts. Some may need limited guidance on classroom management strategies because of a 'flair' which enables them to develop early positive relationships with students. The focus then shifts to monitoring that they accurately observe the practice of others and reflect effectively on their own practice to ensure that they can identify how and why this practice happens in this way (Sfard 2005). Similarly, the particular strength of a pre-service teacher may be a depth of understanding about the purposes and practices of planning for and assessing learning. For such a pre-service teacher, mentoring becomes directed towards effective enactment of planning and resourcing and the skilled use of assessment in planning forward for classes and individuals. In either of these situations, and any other that may be presented by a talented pre-service teacher, the mentoring task becomes a focus on how best to enable these novice teachers to utilise their positive relationships with students and depth of knowledge about teaching-related matters to enhance their students' learning from an early stage in their training.

The very nature of 'talent' in teaching is personal, individual and rarely replicated with any degree of certainty. Thus, working with talented beginning teachers is subject to the very specific context in which the mentor and mentee find themselves. We argue that mentoring a talented beginning teacher is a different experience from mentoring one who is not, since the established patterns and regularities in mentoring which one calls upon are less suitable. In order to support, in a general way, how the mentoring of such beginning teachers can be effective and sustained, we offer three ways of viewing this mentoring process: a theoretical framework, a framework of characteristics of stages of teacher development from observed teaching, and our own experience of *critical moments* in our mentoring of talented pre-service teachers. The first of these, the theoretical framework, presents a model of ways of learning which may resonate with a particular situation or a particular beginning teacher. The second, the characteristics of stages of teacher development, may be easily recognisable in the practice of the beginning teacher and provide a framework for proactively moving the beginning teacher forward. The *critical moments* we describe may trigger identification of similar – or quite varying – context-specific scenarios which can be utilised to promote suitably tailored specific learning for the pre-service teacher.

A theoretical framework

Socio-constructivist learning has long been associated with students' learning and much research has been undertaken in the learning of mathematics within this theoretical model. It is based on an assumption that cognitive development in mathematics, or mathematical learning, is a social, communicative practice. We use this theoretical approach to provide a way of thinking about a talented beginning teacher's learning about teaching mathematics.

Social constructivist ways of working

The importance of developing a shared language for effective collaboration for mentoring is established in Chapter 6, as is allowing a beginning teacher to develop their own identity in relation to both mathematics and to teaching. For social constructivists, no separation can be made between the individual and the social context within which knowledge and understanding develop. Wegerif (2008) argues that there is evidence in dialogue that identity of every sort is constructed. Dialogue is dynamic and directly affected by the environment and is equally applicable to developing a mentor's identity as it is for developing a beginning teacher's identity as they act as participants in dialogue. As Coles (1995: 165) describes, "The social interactions stimulate members of the group to think together; from a psychological point of view this pushes forward the level of thinking and 'scaffolds' his or her cognitive processes".

Mercer (1995) describes what he calls the 'guided construction of knowledge' in which

> people appear to establish some shared understanding . . . someone asserts some intellectual authority during the event . . . all of them represent the joint pursuit of some kind of learning. They all show language being used as a tool, a social mode of thinking, for the development of knowledge and understanding . . . knowledge is neither accumulated nor discovered by learners: it is shaped by people's communicative actions.
>
> (1995: 19)

In this description, the role of the mentor is visible as 'someone who asserts some intellectual authority' and *scaffolding* is also evident in defining learning as being 'shaped by people's communicative actions'. Mercer argues that the creation of shared knowledge and understanding is an active process, rather than a mere pooling of information. Although information might be accumulated, knowledge and understanding is generated by "*working* with information, selecting from it, organising it, arguing for its relevance" (1995: 67). This is the argument for mentors of talented pre-service teachers understanding the need to *work actively* with the 'talent' observed. We acknowledge, as does Mercer, the inevitable power relations between participants in the conversational talk in the mentor–mentee relationship, particularly in the practice-based situation. We propose that, for a mentor of a talented beginning teacher, more equitable relationships might be established which benefit both in terms of mutual learning.

The importance of how the interpersonal relations between dialogue participants affect their learning and therefore their capacity to reason and learn is well documented (Mercer 1995; Mercer & Fisher 1997; Wegerif 2008). Such collaborative work, in which individuals work jointly on the same problem, is linked with ideas such as *scaffolding*, discussed earlier, and an idea developed by the Russian theorist, Vygotsky, termed the *zone of proximal development* (ZPD).

Zone of proximal development

We use this idea of a ZPD to frame how we envisage working with talented beginning teachers. Vygotsky (1978) describes the ZPD as "the distance between the actual developmental level as determined by independent problem-solving and the level of potential development as determined through problem-solving under adult guidance or in collaboration with more capable peers" (1978: 86). While this framework was devised in relation to young learners, we propose that it is equally applicable to novice teachers. In the context of the novice teacher, their ZPD lies beyond what they can achieve on their own and unaided. As the beginning teacher achieves an increasing repertoire of suitable skills and knowledge in their ZPD, these become part of their observable teaching repertoire, thus extending the reach of the ZPD. A difficulty with applying this model in practice is that the potential development of a beginning teacher in the ZPD can only be explained rather than described, because it is identified in relation to others, whereas the skill set already achieved can actually be observed in practice. However, understanding the achievements of talented beginning teachers, and how these are achieved, can be a rich starting point for working alongside them in their movement into their ZPD.

Kinginger (2002) argues that using the ZPD model encourages collaboration and co-authoring in learning. She interprets Vygotsky's model as a complete learning and development situation which encompasses the learning setting and the available resources, including what might be jointly constructed by a mentor and talented beginning teacher. Viewing Vygotsky's ZPD model of learning as a social process promotes a dynamic assessment of potential rather than a static model of assessing potential. Learning, in the context of a beginning teacher, is seen as more than transference of new knowledge from one to another. It is seen as "joint cooperative action" and the ZPD is seen as "an interactive space that holds potential for multiple – and unpredictable – transformations of human identity" (2002: 246). This way of working can therefore have benefits for the mentor as well as the beginning teacher.

Central to Mercer's (1995) theory of a 'guided construction of knowledge' is context and continuity. In this sense, context is taken to mean the broadest interpretation possible for context – beyond the physical setting into the interactions between participants which develop the context. His interpretation of continuity also goes beyond a linear continuous path to mean the fluidity of change and a dynamic interactive flow. "If context and continuity are not well established in a conversation, the thread of a developing joint understanding may be broken and misunderstandings are likely to arise" (1995: 68). Hence, we argue that establishing a strong sense of a shared language about the teaching of mathematics between a mentor and a talented beginning teacher is central not only to ensuring the full development of teaching potential in the novice teacher but also for the mutual development entailed for the mentor. Hagger and McIntyre (1994) advocate that:

> It is difficult to over-emphasise the importance of talking in the process of learning to teach . . . most of the work of mentors with student [pre-service] teachers will be done through talking about teaching. But student [pre-service] teachers will first need to learn how to talk about teaching, in particular how to use the kind of language needed to analyse and evaluate their own teaching carefully and objectively.
>
> (1994: 7)

While this is true of all pre-service teachers, a talented pre-service teacher may already have been enculturated into the language of teaching, perhaps through prior experience as a classroom assistant. The role of the mentor then becomes one of ensuring that the 'labels' associated with particular aspects of the language, particularly evaluative language, are associated correctly with practice and experience.

Some researchers suggest that the context in which learners learn is as important as the means through which learning happens. Lerman (2001) claims that teachers' learning about teaching occurs through increasing engagement with practices that develop personal identities within a community of practice. Such an approach is deemed socio-cultural because it emphasises "the socially and culturally situated nature of mathematical activity, and views learning as a process of appropriating the cultural tools – language structures, symbol systems, forms of representation, and structures of reasoning – recognised by a community of practice" (Goos 1999: 5–6). Within a community of practice, such as that defined by a talented beginning teacher, a mentor and other colleagues in a practice-based setting, *scaffolding* is an important structural feature of providing the beginning teacher with the necessary assistance, activity or experience to enable them to move beyond their zone of achievement into their zone of proximal development.

Scaffolding can take a variety of forms within the mentor's repertoire of skills. Just as for good classroom teaching, pitching the learning appropriately is crucial, so that this takes the form of suitable challenges for a talented beginning teacher which are achievable within their ZPD. For example, a target of developing a classroom presence which is all-pervasive might be scaffolded by organising for the talented beginning teacher to observe a particular teacher – perhaps a drama teacher in a secondary school, or a teacher in a primary school who communicates effectively using very few explicit verbal instructions. In order for a talented beginning teacher to contribute effectively to medium- or long-term planning with colleagues, a mentor might suggest that they read a section of a research-based book on mathematics teaching, such as the chapter 'Connecting decimals and measurement' (Watson, Jones & Pratt 2013), so that they bring experience informed by research instead of their limited teaching experience to the joint undertaking. In this way, the mentor is scaffolding the talented beginning teacher's experience of contributing as an equal to a joint endeavour, experiencing the sense of achievement, and centrally, rather than peripherally, engaging the beginning teacher in a community of practice.

In embracing the socio-cultural approach to learning, we use an extension of Vygotsky's model of a ZPD to Valsiner's (1997) use of a *zone of free movement* (ZFM) and a *zone of promoted action* (ZPA). The *zone of free movement* represents all the affordances and constraints acting on the learning of a pre-service teacher in the context of the practice-based setting. These might include the attainment of the students being taught, degree of adherence to a particular curriculum or scheme of work in a teaching setting or the availability of resources, especially technological. The *zone of promoted action*, or active support for a pre-service teacher, is represented in two fields. One field is the opportunities offered and actively encouraged by the practice-based setting. The other is the expectations of activities required by the provider institution. In an ideal setting, the *zone of promoted action* from both the practice-based setting and the provider institution will overlap to maximum extent, and these will also have maximum overlap with the pre-service teacher's ZPD. Additionally, in an ideal setting, all of the *zone of promoted action* and ZPD will be within the affordances of the *zone of free movement*, as indicated in Figure 8.1.

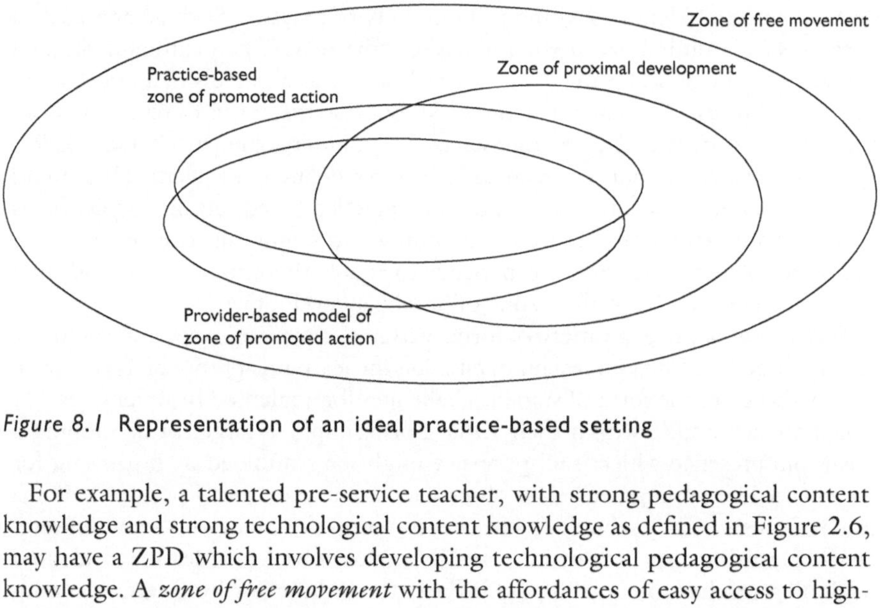

Figure 8.1 **Representation of an ideal practice-based setting**

For example, a talented pre-service teacher, with strong pedagogical content knowledge and strong technological content knowledge as defined in Figure 2.6, may have a ZPD which involves developing technological pedagogical content knowledge. A *zone of free movement* with the affordances of easy access to high-quality technological resources and appropriate software resources, and classroom climates and environments which allow a pre-service teacher to work constructively with students, will provide a positive enabling of this pre-service teacher. The practice-based ZPA will include the extent to which the mentor challenges and encourages the talented pre-service teacher to not only actively take the opportunities on offer but also to take pedagogical risks in availing themselves of these opportunities, so that they are 'testing' the limits of both the *zone of free movement* and their ZPD.

Conversely, we might imagine a different scenario for the same talented pre-service teacher. Within the *zone of free movement*, the constraints in allowing this pre-service teacher to develop within their ZPD might outweigh the affordances. There may be little technology available and even less appropriate mathematical software. The requirements in the *zone of promoted action* from the provider institution, such as teaching a mathematical topic using specific software such as a spreadsheet, for example, may be unattainable for the pre-service teacher. The practice-based setting may offer a limited *zone of promoted action* in the context of this aspect of the pre-service teacher's ZPD because of the limited resources and, possibly, the lack of ease with which students engage with technology-based learning. In this case, the model of engagement within the *zone of free movement* might look like Figure 8.2.

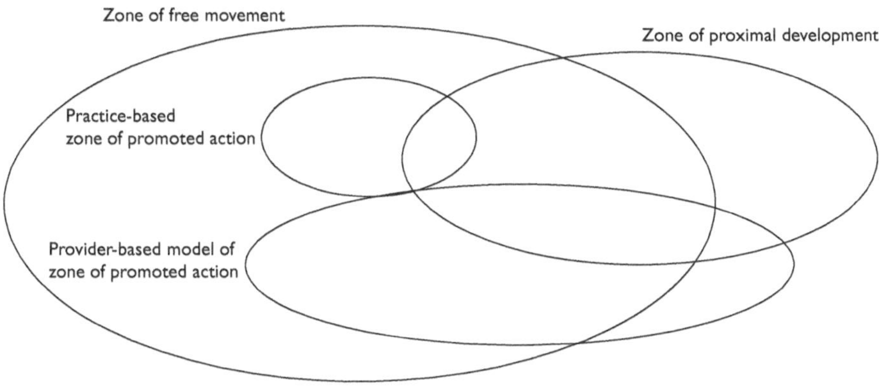

Figure 8.2 Representation of a limited practice-based setting

On the part of the provider institution, matching the ZPD needs of a talented pre-service teacher with the *zone of free movement* and *zone of promoted action* in a practice-based setting is of prime importance. However, logistics often determine that the ideal is not always achieved. Hence the mentor, as a mediator between the *zone of free movement*, the *zone of promoted action* from the provider institution and the pre-service teacher's ZPD, requires a skilled 'eye' and sound listening skills to engage in a *zone of promoted action* with the 'talent' of the pre-service teacher, particularly where the *zone of free movement* is constraining (Goos 1999, 2005). The mentoring role may need to include altering, in some form, the *zone of free movement* for the pre-service teacher. In the case of the talented pre-service teacher described earlier, who is developing technological pedagogical content knowledge, the mentor may need to negotiate teaching experiences in a local practice-based setting which is rich in technology. Analysing the needs of a specific talented beginning teacher and evaluating how to engage in a *zone of promoted action* might effectively be done by recognising where this beginning teacher may lie on a continuum of teaching expertise.

Novice to expert: the characteristics

In order to understand and be able to determine an approach to mentoring a talented beginning teacher, knowledge of understandings of the characteristics of the range of expertise from novice to expert is likely to be helpful. We draw on the work of Berliner (1994) who describes five stages of development of practitioners: *novice*, *advanced beginner*, *competent performer*, *proficient practitioner*, and *expert*. While we acknowledge that stages of development of teaching are rarely linear, outlining some of the features which characterise each of these stages can be helpful in enabling identifying, reflecting on and engaging with specific features.

Novice teacher

At this stage, novice teachers are learning to identify and 'label' components of the practice of teaching, such as 'closed and open-ended questioning', 'positive reinforcement' and 'emotional literacy'. This will happen mainly through targeted observation of experienced teachers with the purpose of the novice teacher realising that these generic components of good teaching are not context-dependent but underpin the *craft knowledge* of effective teaching. The advice given to novice teachers is often at the level of, "Always use positive reinforcement more often than sanctions to manage low level disruption" or, "Make sure that you have some differentiated mathematics material for the children in Green group". At this point, the novice teacher is gathering a set of skills and knowledge, which is rationally determined, and usually procedurally practised as a set of 'rules'. Practice of these components in mathematics teaching tends to be in isolation of each other, as a set of collected skills. Berliner argues that the novice teacher relies largely on previous beliefs and experiences, as identified in Chapter 3, rather than the situated mathematics classroom context in which they are observing and teaching. The novice teacher sees the responsibility for teaching actions as remaining with the 'labels' or rules which are being practised. They can easily work with pre-determined plans for teaching mathematics lessons.

Advanced beginner

Advanced beginning teachers integrate their increasing teaching experience with the 'labels' and advice previously acquired to begin to relate practice to the classroom context in which teaching is happening. They learn when to conform to advice and when to deviate from this as they make judgements about their particular teaching setting. This strategic knowledge, however, is still determined by the framework of 'rules' identified and used initially as a novice teacher. For the examples above, advanced beginning teachers understand when positive reinforcement may not be as influential for managing students' behaviour as using low-level sanctions. Similarly, they recognise that always praising and positively encouraging the most mathematically able students, despite their under-achievement, may not be as productive in encouraging higher-level achievement as a criticism of their performance. So the differentiation in teacher behaviour towards groups or individuals becomes a feature of an advanced beginning teacher's practice. However, they are also still at a stage where they may be content to accept and enact pre-planned lessons. Berliner suggests that, while advanced beginning teachers may be integrating aspects of teaching experience with teacher knowledge of skills and 'rules', they are still not actively assuming responsibility for mathematics teaching activity. He argues that, while advanced beginning teachers are recognising and classifying specific teaching situations in order to act differently, they are not assuming responsibility actively for their decisions, rather ascribing them to the 'labels' and 'rules' of teaching.

Competent performer

This stage is what is sometimes called 'thinking on one's feet'. Competent performers begin to rely on previous experience to determine how they choose to teach in the classroom. They are also characterised by a tendency to plan or re-plan lessons so that they take 'ownership' of mathematical learning aims for the lesson and the means by which they want to achieve these. Competent performers usually decide what paths to follow in the classroom and which to ignore, so maintaining a balance between knowing which are the key aspects of a mathematical topic that need communicating and which to ignore, given any specific context or group of students. Unlike the previous two stages, a competent performer feels more in control of what they are doing and how they do this. Thus, they feel a greater sense of responsibility for what occurs in their classrooms. This sense of responsibility, in turn, Berliner argues, demands a greater emotional attachment to successes and failures in their teaching or their students' learning. This lack of detachment from successes and failures means that these may disproportionately 'loom large' in the consciousness of the competent performer.

Proficient practitioner

This stage, Berliner argues, is when "intuition and know-how become prominent" (1994: 7). As experience increases, reflection on particular classes, particular students, and particular teaching and learning settings provides the proficient practitioner with a range of scenarios which can be identified and linked. Teaching is then seen as a more holistic event, rather than compartmentalised into events which are less familiar (and 'labelled' by the novice teacher). A range of experiences provides the proficient practitioner with a 'sense' of commonalities which is then used to predict situations and proactively manage these when they arise in the classroom.

Expert

The final stage in Berliner's model of expertise is the *expert*. He argues that these experts rely almost wholly on intuition, so that analysis of any situation and the determination of an appropriate response is not usually undertaken. Responses are integral and 'fluid' to the practice of an expert teacher. This fluidity often ensures that teaching happens smoothly. However, the expert can respond to unexpected situations rapidly by drawing on analytic skills and determining an appropriate response. To an observer, this may also appear to be part of the 'fluid' process of classroom teaching.

Implications for mentoring a talented beginning teacher

It is argued in Chapter 7 that "the methods by which mentors support pre-service teachers' reflections will vary according to their stage of training". A talented beginning teacher is most likely to appear to be working at the *competent performer* stage in the classroom, demonstrating aspects of 'thinking on one's feet' from an early stage. However, although this may be the case, the mentor of such a beginning teacher needs to recognise that their lack of experience in many aspects of teaching may mean that they do not necessarily interpret classroom scenarios with the same 'eye' as a *competent performer*. The mentor can increase the rate at which more accurate interpretations occur by providing, in quick succession, a series of activities or tasks which allow the talented pre-service teacher to develop an understanding of patterns in similarities in experiences and so move into the range of the *proficient practitioner*. For example, for a target of improving the 'pitch' of differentiated mathematics activities in a primary classroom, a mentor might suggest that the talented pre-service teacher does the following seven related tasks over three days:

- on day one, examine the curriculum or scheme of work appropriate to the series of lessons;
- on day one, work as a teaching assistant in a mathematics lesson, planned and taught by the mentor, with the specific focus of talking with five identified students in order to understand their thinking as they undertake the mathematics activity assigned to them; write a brief reflective account on 'pitch' relating the curriculum to the conversations with students; discuss this with the mentor;
- later on day one, reflecting on these conversations and discussion with the mentor, plan and choose differentiated activities for the mathematics lesson for the following day;
- on day two, again work as a teaching assistant in the mathematics lesson, planned by the pre-service teacher but taught by the mentor, with the specific focus of talking with the same five students in order to understand their thinking as they undertake the mathematics activity assigned to them; write

a brief reflective account on 'pitch' relating the curriculum to the conversations with students; discuss this with the mentor;
- later on day two, reflecting further on the conversations with students and discussion with the mentor, plan and choose differentiated activities for the mathematics lesson for the following day;
- on day three, teach the mathematics lesson; using knowledge of the five students who have been the focus of the differentiation, observe these students while working with the whole class and talk with them, as appropriate; mentor undertakes a formal observation with a focus on 'pitch';
- later on day three, debrief and discussion of observation with mentor, focusing on 'pitch' and targeting further development.

The importance of challenge in a trusting dialogue between the mentor and beginning teacher cannot be underestimated in this context. As mentioned earlier, if the context and continuity of dialogues are not embedded in conversations, misunderstandings are likely to occur, so trust is an important component of activating challenge.

Encouraging a talented beginning teacher to 'routinise' their actions in the classroom as soon as possible will provide them with a sound footing on which they can build efficiency into teaching, and thus into learning outcomes for students. Again, this comes from ensuring that foci for observation allow the beginning teacher to 'see' the analysis and deliberate actions inherent in the intuitive practice of *expert* teachers, such as clear and established signals (often non-verbal) to start a lesson and move between transitions, use of language when questioning students which efficiently elicits required information, informal (often unnoticed) assessment practices, and effective use of pace. Increasing the rate at which talented beginning teachers 'routinise' actions, and at which they develop the depth of understanding of actions in the expert teaching of others, will help them to move more quickly from *advanced beginner* and *competent performer* stages into the *proficient practitioner* stage.

What is often underestimated in the training of talented beginning teachers is the extent of emotional involvement in their early teaching because they are at a *competent performer* stage. These emotions can be expressed in many different forms for both failures and successes. In particular, emotional responses to failure are seen to be particularly damaging to progression in teaching knowledge and skills. All these emotions need to be carefully managed by the mentor in order to maintain the stability necessary and focus required on 'seeing' similarities in actions and outcomes of specific practices in moving the talented beginning teacher to a *proficient practitioner*. Evidence from analysing teacher behaviour by Blanton and colleagues (2005) provides another feature of teacher action. This occurs when teachers are faced with a challenging situation and choose to revert to more established and familiar routines, rather than develop enhanced responses to a situation. Identifying these for the talented beginning teacher and making these overt for them will enable them to use them as a focus for improvement.

Using critical incidents

It is argued that mathematically gifted students require appropriate intellectual challenges as well as experiences which also enhance and enrich attitudinal and motivational aspects of engaging with mathematics. Similarly, talented beginning teachers require intellectual challenges which provide a focus for analysis, evaluation and reflection. Orland-Barak and Yinon (2005) claim that critical incidents in a talented beginning teacher's experience are essential to this reflective process: "professional learning frameworks designed around the writing, sharing, and reflection on participants' critical incidents, seem to propose an effective channel for exploring the situated and intricate process of acquiring expertise" (2005: 575). We define critical incidents as significant events or circumstances that encourage reflection and provide the observer with an insight into learning and teaching. Lerman (1994) also argues that an analysis of critical incidents is a useful tool in mentoring beginning teachers. He suggests, however, that beginning teachers may need some guidance from mentors in order to help them understand how to 'notice' critical incidents, and how to use them to inform future teaching. We provide some examples from our own experience of critical moments for talented pre-service teachers in their practice-based placements These pre-service teachers are identified by pseudonyms.

Anne: Confidence in decision-making

Although in the final stages of training, Anne had expressed an interest in improving her behaviour management strategies. While she had a good level of control of her classes, she felt she could benefit from exploring a range of strategies to manage both low-level disruption, and more challenging behaviour. As a starting point, Anne's mentor suggested that she observed a range of teachers and classes, paying particular attention to strategies employed to manage behaviour, and identifying specific and potentially key actions by the teacher. This formed the basis for discussion between the mentor and mentee. The observations had offered Anne a broad range of approaches, some which she felt she could adopt and others that she felt did not match her style of teaching. Reflecting on these observations, Anne described how observing expert teachers at this stage in her training was extremely valuable. These observations contrasted with those that took place in the early stages of her teaching, as she was able to relate classroom occurrences to her own teaching experience. This experience enabled Anne to identify patterns and similarities in her experience and that of others and utilise these in developing more rapidly towards a *proficient practitioner.* Additionally, posing fictitious scenarios allowed the mentor to facilitate further discussion about patterns in Anne's developing practice for these circumstances.

A further critical moment in Anne's development as a teacher involved an incident with a student. Anne had given a student a detention for a behaviour-related occurrence in the lesson and had independently made the decision that she would ask him to sharpen a box of pencils as a sanction. When the student defiantly refused and his behaviour deteriorated, she was forced to consider a way forward. Anne's mentor was present during this critical moment and observed Anne's reaction. Confident that Anne would handle the situation appropriately, she did not intervene; instead, she simply offered encouraging nods in Anne's direction, allowing Anne to make her own decisions. In discussion of this critical moment after the event, Anne explained how she had felt uncertain about the appropriate way forward, as she was stuck between a potentially unreasonable sanction and appearing to 'back down'. She described how this incident had not only given her confidence in independently making her own decisions, but emphasised, for her, the importance of setting realistic and achievable expectations and consequences. In doing so, she overtly linked what she had read as an expectation of the provider institution with her first-hand experience in the classroom – an example of how matching the *zone of promoted action* from both the provider institution and the mentor with the beginning teacher's ZPD can provide significant experiences.

Louis: Flexibility in teaching

Louis was a talented pre-service teacher who regularly achieved outstanding gradings for his observed lessons. He experienced a critical moment when he tried to experiment with being more flexible in his delivery of lessons:

> One of my strengths had been having [Powerpoint] presentations prepared for each lesson; the structure of all my lessons had hung on a [Powerpoint] presentation of some kind. We [the mentor and he] discussed how real-life teaching isn't like that, so I should experiment with introducing more flexibility. I wrote out on lined paper my objectives, major points, resources and questions, with no [Powerpoint] presentation. After delivering the lesson I wrote "a process I did not enjoy! My new plan in terms of becoming more flexible is to play to my strengths but to introduce elements of flexibility into 'some' lessons, 'gradually', rather than moving completely away from my usual style of planning."

This critical incident reflects two phases of learning. First, Louis realised that this gradual transformation was what his mentor intended in the first place but reflected positively on the experience. While he did move away from highly structured Powerpoint presentations, this happened gradually, and the experience made him realise which are the more important elements of structure in a lesson and which can be removed. Second, the mentor understood the importance of the clarity of the suggestions she made to a mentee and reflected on how she might manage a different talented beginning teacher in a similar situation. The scenario indicates how a critical moment such as this can aid a talented pre-service teacher to overtly exhibit skills in the *competent performer* stage and allow him to reflect on these, while also providing a learning situation for the mentor.

At a later point, Louis' mentor told him that he had met the standards to qualify as a teacher. While observations of his teaching continued, Louis was not as focused on these as on other aspects of his development. In the confidence of the knowledge that he had 'passed' as a teacher, Louis began to take risks with a challenging class of students.

> Differentiation, behaviour and engagement were all challenging with this class, and spending those extra weeks with them, without having to think about being assessed in an observation, was really helpful. It gave me valuable practice with a challenging class and it definitely helped me build on the progress I had made. This gave me freedom to experiment which has been put to good use in my NQT [newly qualified teacher] year!

The choice made by Louis's mentor to reassure him that he had 'passed' his qualification was a critical moment in refocusing Louis's direction for development as a teacher. The mentor's knowledge of Louis and his determination to improve his teaching had enabled her to make a decisive contribution to his development through a single act.

Final remarks

We have assumed a premise that all pre-service teachers should be equitably encouraged and guided to achieve a teaching qualification at the highest level possible. This is no less the case for talented pre-service teachers. In the interests of providing the richest learning settings for students' learning, developing talented beginning teachers, post-qualification, is equally important. We have argued that, when working with talented beginning teachers, an honest and open acknowledgement of the situation in which both the mentor and novice teacher find themselves is likely to be the first step towards establishing a positive and trusting mentoring relationship. Similarly, acknowledgement of the potential talent of the beginning teacher needs to be overt so that it becomes a focus for reflection and meta-reflection. The development of talented beginning teachers is, because of the very nature of 'talent', likely to be highly individual, not conforming to patterns of usual development. Therefore, responses to a talented beginning teacher need to be similarly personalised and responsive. We propose that exploring ideas such as the *zone of proximal development*, *zone of free movement* and *zone of promoted action* can facilitate mutual engagement in a learning process for both the beginning teacher and the mentor.

The skill of tailoring experiences for a talented beginning teacher is one that is learned through reflection on experiences of mentoring. These, as Orland-Barak and Yinon (2005) argue, are different skills than those of an *expert* teacher. It is not sufficient, in moulding 'talent' efficiently, to be an *expert* teacher oneself. These authors suggest that learning mentoring skills can parallel learning teaching skills. They contend that teaching 'experience' does not necessarily equate to mentoring 'expertise'. Mentors can be challenged by atypical events presented by a talented beginning teacher and experience an, often distressing, sense of reversion to a novice situation. This is because they do not possess the intuitive automation of mentoring *craft knowledge* which is paralleled in their *expert* teaching *craft knowledge.* The frequency of a mentor's experience of a talented beginning teacher is such that the same patterns and regularities of action are not established. However, Orland-Barak and Yinon found that, "Positioning themselves as experts while introspecting in regard to the position of distress, they [mentors] were able to develop meaningful patterns of reorganisation, reinterpretation and representation of the problem more qualitatively" (2005: 573). The process of reflecting on emotions and unexpected events increases the mentor's capacity to identify specific targets, activities, teaching-related tasks and teaching settings appropriately for a talented beginning teacher.

Our use of critical moments is aimed at representing how mentors can scaffold a talented pre-service teacher's ability to 'see' through the expertise of teachers and into the craft knowledge. These critical moments may be unintended, at the time. What makes them critical is the actions talented pre-service teachers take as an outcome of the critical moment and the reflecting that mentors undertake on their actions as a mentor (Orland-Barak & Yinon 2005). A talented beginning teacher is likely to be within the scope of Berliner's (1994) competent performer. What is therefore essential to the task of mentoring a talented beginning teacher is providing a climate of challenge within a community of practice which acknowledges the emotional 'swings and roundabouts' of this beginning teacher. A supportive community of practice will offer the depth and breadth required to enable a talented beginning teacher to take the necessary risks to achieve their potential.

Acknowledgement

We would like to thank Professor John Monaghan from the University of Leeds for conversations with us which have clarified our thinking about the relationships between the ZPD, the *zone of free movement* and the *zone of promoted action.*

References

Berliner, D. (1994) *Mentoring in Primary Schools. Teacher Expertise: Reading 4*, Milton Keynes: Open University Press.

Blanton, M., Westbrook, S., & Carter, G. (2005) 'Using Valsiner's zone theory to interpret teaching practices in mathematics and science classrooms', *Journal of Mathematics Teacher Education*, 8, 5–33.

Coles, M. (1995) 'Critical thinking, talk and a community of enquiry in the primary school', *Language and Education*, 9, 161–77.

Goos, M. (1999) 'Scaffolds for learning: A sociocultural approach to reforming mathematics teaching and teacher education', *Mathematics Teacher Education and Development*, 1, 4–21.

Goos, M. (2005) 'A sociocultural analysis of the development of pre-service and beginning teachers' pedagogical identities as users of technology', *Journal of Mathematics Teacher Education*, 8, 35–59.

Hagger, H. & McIntyre, D. (1994) *Mentoring in Primary Schools. Observing experienced teachers: Reading 5*, Milton Keynes: Open University Press.

Kinginger, C. (2002) 'Defining the zone of proximal development in US foreign language education', *Applied Linguistics*, 23(2), 240–61.

Lerman, S. (1994) 'Reflective practice', in B. Jaworski & A. Watson (eds) *Mentoring in Mathematics Teaching*, London: The Falmer Press, pp. 52–64.

Lerman, S. (2001) 'Accounting for accounts of learning mathematics: Reading the ZPD in videos and transcripts', in D. Clarke (ed.), *Perspectives on Meaning in Mathematics and Science Classrooms*, Dordrecht: Kluwer Academic Publishers, pp. 53–74.

Mercer, N. (1995) *The Guided Construction of Knowledge*, Clevedon: Multilingual Matters.

Mercer, N. & Fisher, E. (1997) 'The importance of talk', in R. Wegerif & P. Scrimshaw (eds) *Computers and Talk in the Primary Classroom*, Clevedon: Multilingual Matters, pp. 13–37.

Orland-Barak, L. & Yinon, H. (2005) 'Sometimes a novice and sometimes an expert: Mentors' professional expertise as revealed through their stories of critical incidents', *Oxford Review of Education*, 31(4), 557–78.

Sfard, A. (2005) 'Challenging discourse', in A. Chronaki & I. Christiansen (eds) *Challenging Perspectives on Mathematics Classroom Communication*, Greenwich, CT: Information Age Publishing, pp. 323–38.

Valsiner, J. (1997) *Culture and the Development of Children's Action: A theory of human development*, New York: John Wiley and Sons.

Vygotsky, L. (1978) *Mind in Society*, Cambridge, MA: Harvard University Press.

Watson, A., Jones, K., & Pratt, D. (2013) *Key Ideas in Teaching Mathematics*, Oxford: Oxford University Press.

Wegerif, R. (2008) 'Dialogic or dialectic? The significance of ontological assumptions in research on educational dialogue', *British Educational Research Journal*, 34(3), 347–61.

Part 3

Developing a partnership

Chapter 9

Developing the mathematics teacher as mentor through team-based approaches to mentoring*

David Wright and Rachel Lofthouse

This chapter discusses the context for the encounters between mentor, pre-service teacher and supervising tutors and provides an example of how tutors, mentors and pre-service teachers worked together to develop a tool and process to enhance the observation and feedback experienced by pre-service teachers. We describe the experience of pre-service teachers, school-based mentors and supervising tutors as they work at the boundary between the practices of schooling and teacher education and exemplify how a team-based, enquiry-focused approach supports and mediates learning of all three agents through identification, co-ordination, reflection and transformation.

Introduction

> The year after . . . I secured a full-time teaching position at Aramis Park High School . . . There was one other PE teacher at Aramis when I arrived and we both shared a staff room with the History/Geography teachers. They were a very friendly group, and were occasionally given to practical jokes, especially with first year teachers . . . Understandably, I didn't take much notice when my name went up on the board to volunteer as a practicum advisor. Just another joke! Needless to say I was a little surprised when, on the Wednesday before an introductory practicum for student teachers at Aramis High, a young man knocked on the History/Geography staff room door and asked to speak to 'Mr Clarke', his practicum advisor. This was my first year of teaching and here was my first student teacher.
>
> (Clarke 1997: 166)

Although the situation described in this example may seem extreme, we suspect that a significant number of teachers find themselves as mentors with scarcely more preparation. Moreover, it is clear that the position of mentor to a pre-service teacher is a key role in teacher education and the quality of this relationship can

*An earlier version of this chapter was published in *International Journal of Mentoring and Coaching in Education*, 1(2), 89–103 as Lofthouse and Wright (2012).

have a significant impact on the professional learning of both the pre-service teacher and mentor.

Our premise is that in training programmes that emphasise school experience, tutors, pre-service teachers and their mentors are engaged in a process of boundary crossing (Akkerman & Bakker 2011; Engeström 1999). From a situative perspective, the pre-service teachers are moving from peripheral participation in the school practice of teaching to recognised membership of the community of practice of teaching of the school (Wenger 1998). At the same time as the pre-service teacher is moving from being a student with a base at an accredited provider (such as a university) to becoming a teacher, the mentor is crossing a practice boundary, from school teacher to teacher educator. Supervising tutors, too, are involved in this process through which we problematise our practice by a process of enquiry.

Interactions between the pre-service teacher and mentor also often occur at a boundary between the specific practices and expectations of the accrediting provider (often a university) and the school. In practice-based programmes, such as those found in England, pre-service teachers, their supervising tutors and school-based mentors can also find themselves working at a conceptual boundary between the competency-based training and monitoring that is often found in schools and the reflective mode of professional learning that is promoted by accrediting bodies such as universities. Each approach can bring with it artefacts of practice and specific discourses, one highly regulatory, the other more discursive. These boundary crossing situations can be sources of tension, but are also potential sources of learning. For example, Akkerman and Bakker (2011) in their review of literature on boundary crossing, distinguish four potential learning mechanisms which can take place at boundaries: identification, co-ordination, reflection and transformation.

Models of mentoring

One model of initial teacher education, for example that adopted by the current (2012) English government which uses the word 'training', could be characterised by the scenario where the mentor is the 'expert' who guides the 'novice' teacher in their role as they 'learn the ropes'. The novice produces lesson plans under the guidance of the mentor and attempts to carry out activities with learners to support their developing understanding of mathematics. In this rather mechanistic model, learning mathematics and learning to become a teacher are unproblematic activities characterised by a transfer of knowledge from the expert to the learner (the school student or pre-service teacher). Variations on this theme might focus on the relationship between the mentor and pre-service teacher with the mentor interpreting their role slightly differently as, for example, a colleague of the pre-service teacher or the pre-service teacher as an apprentice who learns the 'craft'. In this arrangement, the supervising tutor becomes a quality control figure for both the mentor and pre-service teacher who visits to 'check up' on the progress of the pre-service teachers and who may also have a role in evaluating the quality of mentoring. Underpinning these activities is a model of increasing measurement and accountability in the education system. However, as Edwards, Gilroy and Hartley (2002) argue, apart from its appeal to 'common sense', this approach has little rational or evidential base (2002: 3).

Research into learning mathematics (Hiebert & Carpenter 1992; Nunes, Bryant & Watson 2009) and research into learning to be a teacher (Eraut 1994; Goos 2008; Lave & Wenger 1991; Oliveira & Hannula 2008) challenges these certainties. For example, for at least two decades there has been a consensus among researchers that learning mathematics should be viewed "as both a process of active individual construction and a process of enculturation into the mathematical practices of the wider society" (Cobb 2005: 40). Moreover, it could be argued that teachers of mathematics learn to teach by "building up a repertoire of increasingly informed interpretations of affordances, and responses which make use of those affordances *in the act of teaching*" (Edwards *et al.* 2002: 110). This is a complex situation and in this context the role of the mentor moves from that of a 'trainer' to a more nuanced agent. Indeed, it is clear that there is no straightforward step from an experienced classroom teacher of mathematics to a mentor of pre-service teachers, even though many mentors appear to have been chosen on this basis.

Opportunities for learning

Supervising tutor, mentor and pre-service teacher have different mental maps, different observational skills and different levels of access to knowledge. Some of these are outlined in Chapter 8. Our argument is that the diverse perspectives and practices experienced by pre-service teachers, their mentors and supervising tutors provide opportunities for professional learning through *dialogicality*, drawing on Bakhtin's (1981) philosophy in which learning is a process that involves multiple perspectives and multiple parties where power may be more equally shared between them (Flecha 2000).

Our focus here is on classroom observation and feedback by practice-based mentors and on the use of an observation tool as an artefact to enhance pre-service teachers' and mentors' learning. We argue that in this activity both the mentors and the pre-service teachers engage in developing their professional identity, shaped by sometimes differing practical expectations, academic demands and cultural values of both an accrediting provider (such as a university) and a school. In the case where joint observations are shared with supervising tutors, we also suggest that there is the potential for significant professional development and learning to be triggered by crossing both real and metaphorical boundaries. As researchers, we also believe that the educational design process (Shavelson, Phillips, Towne & Feuer 2003) enables us to engage with practice in a significant development of our role. We believe it is essential that the tools offered to pre-service teachers and their mentors are supportive of divergent learning outcomes, through which each pre-service teacher and mentor has the opportunity to transform teaching practices, not simply replicate existing ones. This is vital to address current challenges in teacher education of transforming learning in a context of rapid change.

There is a growing interest in the development of 'boundary objects' as artefacts which fulfil a bridging function (Edwards, Lunt & Stamou 2010; Edwards & Mutton 2007). We suggest that the activity of classroom observation and feedback by school mentors is a boundary activity where a range of learning opportunities occurs and that artefacts can be designed to enhance these opportunities. We outline the development of an observation tool as a boundary object which mediates the learning undertaken by pre-service teachers and their mentors (and supervising tutors) as they work at the boundary between the practices of schooling and teacher education.

Difficulties with current practice

Pre-service teachers in England experience multiple cycles of observation and feedback by their mentor during their initial teacher education. Such observation absorbs many tens of hours, forms the basis of the mentoring, monitoring and reporting processes that are associated with compliance to programme requirements, and provides the pre-service teachers with their benchmarks of progress. For some pre-service teachers, observation can turn into a purely bureaucratic activity, which does little to engage them or their mentor in anything more than incremental target setting and monitoring. Hargreaves (2000) has characterised such itemisation and categorisation of teachers' work and recognition as "checklists of performance standards or competencies" (Hargreaves 2000: 152). Such a view can affect pre-service teachers' experiences of professional learning in school and their mentors' expectations of their role.

The problem for pre-service teachers is that their initial framing of their classroom experience is likely to be their "default" settings carried with them tacitly (Eraut 2000) from previous expectations, as identified in Chapter 7, and their experiences in the classroom (Barnes 1992). It is difficult to consider alternatives if you have little experience of them.

> Real change is not simply a matter of psychology, of convincing yourself and then acting differently. Real change is based on becoming aware of possibilities which were not previously available . . . at the heart of change is the recognition of new possibilities for acting.
>
> (Mason 2002: 144)

This might also be said to apply to many mentors in terms of their experience of teacher pedagogy. Established practice does little to break this cycle, and this is potentially exacerbated when former pre-service teachers become mentors and reproduce their own experience with their mentees. Evidence from a mixed methods longitudinal study (Tracey *et al.* 2008) shows that observation and feedback practice in English schools rarely supports teachers' self-development as reflective practitioners and tends to undermine any sense of agency on the part of teachers.

Nonetheless, experience suggests that some pre-service teachers find the observation process to be inclusive, reflexive and genuinely developmental for both parties. Our interest as tutors is in promoting this process during teaching placements and in developing our own understanding of it. In doing so, we recognise, along with Edwards and Blake (2007), that even reflective practice is coded within the political and economic contexts within which initial teacher education operates. However, the aim is to foster an approach which goes beyond low-level reflection as part of a technical-rationalist paradigm (Edwards & Blake 2007). The development of a tool which affords the learning opportunities which pre-service teachers might experience by sustained interactions with their mentors and which focuses on their development as teachers is therefore a priority.

Investigating pre-service teachers' experience of observation and feedback

Our investigations (Lofthouse & Wright 2012) suggest a model of differentiated impact of observation and feedback for pre-service teachers. Four levels are identified: (1) at a basic level the pre-service teachers see observations as opportunities to gain feedback and have their progress in meeting specific targets monitored by their mentor; (2) pre-service teachers who appear to have been more activated by the observation and feedback session demonstrate engagement in self-evaluation of the specific lesson; (3) other pre-service teachers recognise a cumulative impact as observations enable them to consider their progress over time alongside their mentor; and (4) at the most sophisticated level, pre-service teachers show how the process has helped to embed reflection into their suite of professional skills, thus increasing their capacity to judge and develop their own practice. These levels are further defined and exemplified below.

A new observation tool

In response to these investigations a new tool for lesson observation was designed (Lofthouse & Wright 2012). As our initial teacher education programme promotes a practitioner enquiry mode of learning, the development of the observation tool was premised in the belief that *questions* provide the key to learning (Blanton, Berenson & Norwood 2001). A question can stimulate the learner's zone of proximal development (Vygotsky 1978), be the trigger for a learning conversation or the basis for an extended enquiry. This further exemplifies the work in Chapter 8. The challenge was to ensure that questioning could be built into the observation and mentoring process, and plays a part in enhancing professional development (Lofthouse *et al.* 2007). The objective was to promote enhanced reflection, as defined by Dewey who proposed that:

> [R]eflective thinking, in distinction from other operations to which we apply the name of thought, involves (1) a state of doubt, hesitation, perplexity, mental difficulty, in which thinking originates, and (2) an act of searching,

> hunting, inquiring, to find material that will resolve the doubt, settle and dispose of the perplexity.
>
> (Dewey 1933: 12)

We were therefore valuing thoughtful tentativeness and a search for evidence.

Thus the new tool promoted questions from both the pre-service teacher and the mentor observer. The link to practitioner enquiry was made as the co-participants owned the questions which were triggered by what was currently going on in their practice which was 'causing some sort of disturbance' (Lofthouse *et al.* 2007: 173). Hence the use of 'enquiry' in Schön's sense where for professional practitioners 'practice is a kind of research' where "their enquiry is a transaction with the situation in which knowing and doing are inseparable" (Schön 1983: 165). Dewey's (1933) influence was also strong as we were challenging a belief that practice and theory are dichotomous, and accepting that "all inquiry is practical, concerned with transforming and evaluating the features of the situations in which we find ourselves" (Hookway 2010: Section 4.2). The hypothesis was that by engaging observer and observed as agents in the same process, both would learn.

Design principles

Three key considerations shaped our practice as tutors and the design of the observation models and the research into their application and impact. First, the belief that a key feature of learning to teach is dealing with inconsistencies, incomplete theoretical frameworks and "situations of uncertainty, instability, uniqueness and value conflict" (Schön 1983: 49). Edwards and Blake (2007) argue that for pre-service teachers "possibilities for change are dependent on the availability of multiple, competing perspectives of teacher identity and practice" (2007: 50). Observations that are based on simple reporting and/or evaluation of practice seem unlikely to support the reflective re-framing of experience needed to deal with these tensions or provide these multiple perspectives (Schön 1983). Second, given the frequency and significance of the observation process in initial teacher education, a pre-service teacher should feel a sense of ownership and shared responsibility for them and the related discussions. If a pre-service teacher is forced to rely on the observer for feedback, and if this is simply framed as an evaluation of competence, it seems likely that their genuine engagement with the process will be put at risk. Thus our research design used a question-based approach to observation to throw back the initiative to the pre-service teacher, putting them in the role of active learner (Blanton *et al.* 2001).

Finally, we theorised that this new observation 'tool' (Jahreie & Ludvigsen 2007; Perks & Prestage 2008) could be an effective boundary object and change the approach taken to observation, making the link between practice evidence and conceptualising classroom processes and outcomes. Akkerman and Bakker's (2011) review claims that "boundary objects have different meanings in different social worlds but at the same time have a structure that is common enough to make them recognizable across these worlds" (2011: 140), hence our conjecture that the observation tool could act as an artefact mediating the social worlds of school and accrediting provider, particularly where a supervising tutor could be involved in a joint observation.

The development of the tool

The development of the tool followed an educational design process. Shavelson *et al.* (2003: 26, cited in Cohen *et al.* 2011) suggest that the key principles of design studies are that they are (a) iterative, (b) process focused, (c) interventionist, (d) collaborative, (e) multileveled, (f) utility oriented, and (g) theory driven. Hence the design of the tool led to an iterative, collaborative, process-focused approach to the development of the observation tool. Pre-service teachers and their mentors were encouraged to experiment with, and report on, their observation experience. Their feedback helped us to define the role of observation in the process of professional learning and to review the nature of the mentoring relationships which emerged. The development process was conducted over three years enabling the tool to be refined with three cohorts of pre-service teachers.

The model to emerge consisted of four stages which were captured on a pro forma (see Figure 9.1).

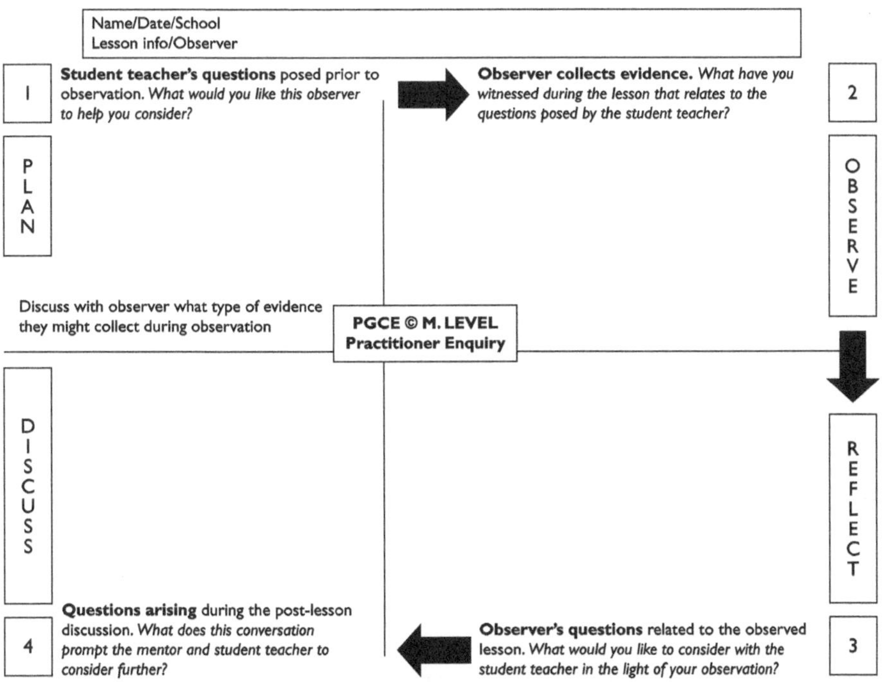

Figure 9.1 Observation pro forma, adapted from Lofthouse and Wright (2012).

Stage 1: pre-lesson

The pre-service teacher records question(s) prior to the observation which they would like the observer to consider during the lesson. Often the first step to reflection is to be aware of a 'disturbance' (Mason 2002), see Chapter 12, and the first stage in our revised observation tool is for the pre-service teacher to identify an issue which has caused some 'disturbance' for them and to pose a question(s) relating to this issue.

Stage 2: gathering evidence

This stage required the observer to document evidence emerging from the lesson that related to the pre-service teacher's question(s). As such, the observer was prompted to carefully analyse the classroom activity and identify evidence which they believed related to the question initially posed by the pre-service teacher. The observer was then able to support the pre-service teacher in 'framing' the situation by providing evidence or a witness statement composed of an account composed of '*brief but vivid*' (Mason 2002: 50) incidents. The intention was to create opportunities for joint reflection on a shared object, i.e. the observation tool, which itself was descriptive, rather than evaluative, and allowed the pre-service teacher to insert their own interpretation into the observation.

Stage 3: posing questions (concurrent with stage 2)

The challenge here is for the observer to transform their evaluative judgements into questions for the pre-service teacher relating to the lesson and the question posed by the pre-service teacher. Clearly, the quality of the questions posed is crucial in providing a stimulus for the novice teacher to reframe their practice and this is a fruitful area for development when we work with mentors.

Stage 4: Post-lesson discussion

The observation debrief is then based on discussing some or all of these questions, and leads to recording of questions that are prompted by this process. This stage involves the mentor and pre-service teacher addressing the issues and questions raised by the questions captured in the first and second stage of the tool. Questions arising from this discussion may then feed forward and provide the first stage of a subsequent observation, form the basis of discussions with other colleagues or pre-service teachers, or perhaps trigger an action enquiry or an investigation of relevant literature or policy.

Discussion

In practice, it was found that the 'tool' shifted the focus of observation to practice as enquiry. In this instance practice can be defined as both the pre-service teacher's classroom practice and the mentoring practice of experienced teachers. While observation of and feedback on classroom practice of pre-service teachers by experienced teachers has become a central feature of the learning landscape of initial teacher education in England it can become a focus for a clash of cultures. The pre-service teacher and their mentor are working at a border between two 'Zones of Promoted Action' (Blanton, Westbrook & Carter 2005; Valsiner 1997), see Chapter 8, that of the school which may be dominated by the performative culture (Ball 2003) and that of the accrediting provider with its more discursive and reflective culture and emphasis on practice as enquiry. The barring of 'judgements' and the encouragement for the observer to reformulate their evaluations as questions was a key element in transforming the relationship and the perspective of the mentor as observer.

Although pre-service teachers and mentors find themselves in these potentially problematic 'borderlands' when discussing classroom practice, the situation can create opportunities for learning. As previously noted, Akkerman and Bakker (2011) argue that boundary crossing situations create the opportunity for dialogic learning as a process involving multiple perspectives and multiple parties and the mechanisms of dialogic learning involve identity, co-ordination, reflection and transformation.

Identity and ownership

"What is typical in identification processes is that the boundaries between practices are encountered and reconstructed, without necessarily overcoming discontinuities. The learning potential resides in a renewed sense making of different practices and related identities" (Akkerman & Bakker 2011: 144). As pre-service teachers and mentors contributed questions for each other's consideration, the observation form was passed back and forth. This process seemed to reinforce both parties' desire to discuss the questions that each had written.

As one pre-service teacher stated, "nobody really led. We would ask questions to each other as we were exploring" and another wrote that he "felt more ownership" and that it "allowed me to have more confidence as it was more of a collaborative discussion with my mentor". In this context, pre-service teachers appeared more willing to initiate discussions around their chosen queries or disturbances, thus enhancing their sense of agency or 'ownership' of the process. These accounts suggest that the tool provided a *dialogical* forum where mentors and pre-service teachers were negotiating their roles in the borderland between school and the accrediting institution. The focus on questions encouraged the mentors to identify the pre-service teacher's learning and their own challenge to formulate effective questions (cf. Bloom's 1956 taxonomy).

Co-ordination

As a 'boundary object' the pro forma serves as a channel of communication and co-ordination between the agents (mentors, pre-service teachers and supervising tutors) participating in teacher education. "The potential in the coordinative mechanism resides not in reconstructing but in overcoming the boundary, in the sense that continuity is established, facilitating future and effortless movement between different sites" (Akkerman & Bakker 2011: 12).

Reflection

"Where identification represents a focus on a renewed sense of practices and a reconstruction of current identity or identities, reflection results in an expanded set of perspectives and thus a new construction of identity that informs future practice" (Akkerman & Bakker 2011: 147). Questionnaire and interview responses revealed that the enquiry-based observation tool gave pre-service teachers opportunities to focus on specific areas of pedagogy with their mentor:

> I think we were trying to get to the bottom of it [explanation]. In a way I don't think we did come up with a conclusion to it, but all the things, the process of going through it is helpful as well in terms of exploring the issue more deeply than you would say in a normal lesson observation; where you would have a number of targets at the end that you would talk about on a more surface level. Whereas this probed further for both of us, and helped M [mentor] think about his explanation, which helped me.

A mentor reinforced this:

> The pro forma questions provide the base for starting the discussion but this process was still open to suggesting further questions from both parties. This meant that the complexities of the issue were revealed and therefore prompted greater student understanding.

Transformation: working on areas of uncertainty or disturbance

"Transformation leads to profound changes in practices, potentially even the creation of a new, in-between practice, sometimes called a boundary practice" (Akkerman & Bakker 2011: 147). The focus on questions seemed to permit pre-service teachers to reflect, discuss and experiment at the boundaries of their professional knowledge or in areas of uncertainty. These types of experiences also created an interesting challenge for the pre-service teachers in relation to the notion of expertise, and as such appeared to reassure them that developing fully comprehensive knowledge and every possible professional skill was unrealistic during their initial teacher education. This led to the recognition that becoming reflective practitioners was what had made a real difference to their achievement as teachers. For mentors and supervising tutors, this process opens up the classroom observation and feedback process as a research tool where the quality of questions becomes a clear priority.

Final remarks

Encouragement for further enquiry is given by Edwards and Mutton (2007) who investigated "how identities are enacted; and systems sustained or adjusted as people work at the boundaries of schools and departments of education in universities" (2007: 504) and provided an extensive account focused on the experience of mentors and conclude that:

> If the tool is to maintain its function as an artefact which brings the two systems together it would seem that it needs to retain, at least on occasions, its position as a boundary object which is jointly worked on at the boundaries of the systems. That is, jointly produced tools, such as assessment instruments, need to be revisited to keep the collaboration alive.
>
> (2007: 509)

The need to 'revisit' corresponds to the iterative nature of design studies (Shavelson *et al.* 2003) and continuing enquiry provides opportunities for revisiting the tool and gathering data from the mentors. Observation can be considered as a boundary activity which exists at a potentially multi-faceted boundary. Both mentors and pre-service teachers are engaged in developing a new professional identity, shaped by sometimes differing practical expectations, academic demands and cultural values of both an accrediting provider (such as a university) and a school. Practices, such as observation and feedback, are critical in defining the experience of pre-service teachers and mentors.

As supervising tutors, we are finding ourselves in a rapidly changing educational landscape and we suggest that this kind of research and design process enables us to actively engage in reshaping our own professional identities. Significant professional development and learning can be triggered by crossing both real and metaphorical boundaries and, as such, it is essential that the tools developed with pre-service teachers and their mentors are supportive of divergent learning outcomes, through which each pre-service teacher and their mentor have the opportunity to transform teaching practices, not simply replicate existing ones. This is vital if, as Hargreaves predicted in 2000, we have entered a post-modern era in which "the context of teaching is changing dramatically, and older modernistic versions of professional and professionalization will not be sufficient to address these significant changes" (Hargreaves 2000: 172).

References

Akkerman, S. F. & Bakker, A. (2011) 'Boundary crossing and boundary objects', *Review of Educational Research*, 81(2), 132–69.

Bakhtin, M. (1981) 'Discourse in the novel' (C. Emerson & M. Holquist, Trans.), in M. Holquist (ed.), *The Dialogical Imagination*, Austin: University of Texas Press, pp. 259–422.

Ball, S. J. (2003) 'The teacher's soul and the terrors of performativity', *Journal of Education Policy*, 18(2), 215–28.

Barnes, D. (1992) 'The significance of teachers' frames for teaching', in T. Russell & H. Munby (eds), *Teachers and Teaching: From classroom to reflection*, London: Falmer, pp. 9–32.

Blanton, M., Westbrook, S., & Carter, G. (2005) 'Using Valsiner's Zone Theory to interpret teaching practices in mathematics and science classrooms', *Journal of Mathematics Teacher Education*, 8, 5–33.

Blanton, M. L., Berenson, S. B., & Norwood, K. S. (2001) 'Exploring a pedagogy for the supervision of prospective mathematics teachers', *Journal of Mathematics Teacher Education*, 4, 177–204.

Bloom, B. (1956) *Taxonomy of Educational Objectives: The classification of educational goals Handbook 1, Cognitive domain*, London: Longmans, Green & Co.

Clarke, A. (1997) 'Advisor as coach', in J. Loughran & T. Russell (eds), *Teaching About Teaching: Purpose, passion and pedagogy in teacher education*, London: RoutledgeFalmer, pp. 164–80.

Cobb, P. (2005) 'Where is the Mind? A coordination of sociocultural and cognitive constructivist perspectives', in C. Fosnot (ed.), *Constructivism: Theory, perspectives and practice*, 2nd edn, New York: Teachers College Press, pp. 39–57.

Cohen, L., Manion, L., & Morrison, K. (2011) *Research Methods in Education*, 7th edn, Abingdon: Routledge.

Dewey, J. (1933) *How We Think: A restatement of the relation of reflective thinking to the educative process*, Boston: D.C. Heath.

Edwards, A. & Mutton, T. (2007) 'Looking forward: rethinking professional learning through partnership arrangements in Initial Teacher Education', *Oxford Review of Education*, 33(4), 503–19.

Edwards, A., Gilroy, P., & Hartley, D. (2002) *Rethinking Teacher Education: Collaborative responses to uncertainty*, London: RoutledgeFalmer.

Edwards, A., Lunt, I., & Stamou, E. (2010) 'Inter-professional work and expertise: New roles at the boundaries of schools', *British Educational Research Journal*, 36, 27–45.

Edwards, G. & Blake, A. (2007) 'Disciplining the practice of creative inquiry: the suppression of difference in teacher education', *International Journal of Research & Method in Education*, 30(1), 33–55.

Engeström, Y. (1999) 'Activity theory and individual and social transformation', in Y. Engeström *et al.* (eds), *Perspectives on Activity Theory*, Cambridge: Cambridge University Press, pp. 19–38.

Eraut, M. (1994) *Developing Professional Knowledge and Competence*, London: The Falmer Press.

Eraut, M. (2000) 'Non-formal learning and tacit knowledge in professional work', *British Journal of Educational Psychology*, 70, 113–36.

Flecha, R. (2000) *Sharing Words. Theory and practice of dialogic learning*, Lanham: Rowman & Littlefield.

Goos, M. (2008) 'Sociocultural perspectives on learning to teach mathematics', in B. Jaworski & T. Wood (eds) *The International Handbook of Mathematics Teacher Education Volume 4: The mathematics teacher educator as a developing professional*, Rotterdam: Sense Publishers, pp. 75–91.

Hargreaves, A. (2000) 'Four ages of professionalism and professional learning', *Teachers and Teaching: History and Practice*, 6(2), 151–82.

Hiebert, J. & Carpenter, T. (1992) 'Learning and teaching with understanding', in D. Grouws (ed.), *Handbook of Research on Mathematics Teaching and Learning*, New York: Macmillan, pp. 65–97.

Hookway, C. (2010) 'Pragmatism', *The Stanford Encyclopedia of Philosophy*. Online. Available at HTTP: <http://plato.stanford.edu/> (accessed 20 February 2013).

Jahreie, C. & Ludvigsen, S. (2007) 'Portfolios as boundary object: learning and change in teacher education', *Research and Practice in Technology Enhanced Learning*, 2(3), 299–318.

Jaworski, B. & Wood, T. (eds), *The International Handbook of Mathematics Teacher Education Volume 4: The Mathematics Teacher Educator as a Developing Professional*, Rotterdam: Sense Publishers, pp. 75–91.

Lave, J. & Wenger, E. (1991) *Situated Learning: Legitimate peripheral participation*, Cambridge: Cambridge University Press.

Lofthouse, R. & Wright, D. (2012) 'Teacher education lesson observation as boundary crossing', *International Journal of Mentoring and Coaching in Education*, 1(2), 89–103.

Lofthouse, R., Wright, D., Riches, A. and Sellers, E. (2007) 'A new model of observing students teach: an opportunity for practitioner enquiry', 3rd Escalate ITE Conference, Cumbria University, Lancaster.

Mason, J. (2002) *Researching Your Own Practice: the discipline of noticing*, London: RoutledgeFalmer.

Nunes, T., Bryant, P., & Watson, A. (2009) *Key Understandings in Mathematics Learning*, London: Nuffield.

Oliveira, H. & Hannula, M. (2008) 'Individual prospective mathematics teachers: studies on their professional growth', in K. Krainer & T. Wood (eds.), *The International Handbook of Mathematics Teacher Education Volume 3: Participants in Mathematics Teacher Education*, Rotterdam: Sense Publishers, pp. 13–34.

Perks, P. & Prestage, S. (2008) 'Tools for learning about teaching and learning', in B. Jaworski & T. Wood (eds.), *The Mathematics Teacher Educator as a Developing Professional*, Rotterdam: Sense Publishers, pp. 265–80.

Schön, D. (1983) *The Reflective Practitioner: How professionals think in action*, London: Basic Books.

Shavelson, R. J., Phillips, D. C., Towne, L., & Feuer, M. J. (2003) 'On the science of education design studies', *Educational Researcher*, 32(1), 25–28.

Tracey, L., Homer, M., Mitchell, N., Malerez, A., Hobson, A. J., Ashby, P., & Pell, G. (2008) *Teachers' Experiences of Their Second Year in Post: Findings from phase IV of the Becoming a Teacher project*, Nottingham: DCSF.

Valsiner, J. (1997) *Culture and the Development of Children's Actions: A theory of human development*, 2nd edn, Chichester: John Wiley & Sons.

Vygotsky, L. (1978) *Mind in Society*, Cambridge, MA: Harvard University Press.

Wenger, E. (1998) *Communities of Practice: Learning, meaning, and identity*, Cambridge: Cambridge University Press.

Chapter 10

The pre-service mathematics teacher as an adult learner

Mark Simmons and Jenni Back

Teaching adults to teach differs significantly from teaching learners to do mathematics. All adults have experience of teaching practices, having been on the receiving end of them themselves. In this chapter, we explore ways that practice-based mentors can direct pre-service teachers towards experiences and models appropriate to the learning environment and their developmental needs. We use Lave and Wenger's apprenticeship model and Schön's reflective practitioner model to support practice-based mentors' thinking about how pre-service teachers are not only learning a body of knowledge about teaching, but also developing a complex and difficult set of skills.

Introduction

Teaching mathematics to primary or secondary school-age students and supporting pre-service teachers in learning to become teachers of mathematics are both tasks for mentors. Some highly skilled classroom mathematics teachers are not effective mentors. In this chapter we identify some of the ways in which supporting pre-service teachers is different to teaching younger learners and some of the attitudes and approaches which will be helpful to mentors aiming to bring out the very best in their mentees. We consider how to maximise the learning from pre-service teachers' interactions with their mentor, their peers who are also pre-service teachers, their supervising tutors and colleagues in the mathematics department, year team and school. Finally, we try to identify many of the key issues which can impinge upon the learning of pre-service teachers in schools and ways to help manage the effects of these.

Models of professional learning

In trying to understand the ways in which adults can learn to become members of a skilled and knowledgeable profession we draw on the ideas of a small number of theorists who have put forward useful models and explanations over the years. Jean Lave and Etienne Wenger (1991) have described positions, such as that of the pre-service teacher, as being one of a 'legitimate peripheral participant' in the work of the 'community of practice' (here we mean the department, year team, school and ultimately teaching in general). The pre-service teacher will pose naive questions of their hosting colleagues, especially in the early stages of a school placement, which is to be both expected and encouraged.

'Communities of practice' theory is often quoted but rarely well developed in its application. It is frequently applied to arenas where the relationship between the 'novice' and 'expert' participants is only loosely analogous to the setting in relation to which the concept was developed. In the case of the pre-service teacher working in a school, the setting and roles closely mirror the original context that forms the case study for Wenger's (1998) seminal work, which was the claims department in an insurance company. Trainee clerks in his study were learning to process claims and, although there was significant documentation around the process available to them, a key source of information as they learnt to operate in this new community of practice was their mentors who demonstrated how to act and offered them advice and role models for how to operate in this context. School settings with the pre-service teacher learning, both indirectly and directly, from other teachers have many similarities with this.

Central to Wenger's theory are notions of practice, meaning and community. His theory is bound up with the notion of human beings as social animals who act interpersonally to make sense of the world for themselves, and it is this process of making sense which constitutes meaning, rather than meaning as it might be captured in the pages of a dictionary. Wenger suggests that the act of making sense for oneself of new contexts and different scenarios involves both thought and interaction with others. When individuals are exposed to new settings, or old settings in which they take on a new role, their capacity to make sense is stretched to its limit and their relationships with others in the context may be under threat.

The notion of developing pre-service teachers as reflective practitioners also has relevance in this context. Donald Schön (1983) coined the term 'reflective practitioner' with the purpose of explaining a learning process in which complex skills and practices cannot be comprehended in a purely theoretical form. Pre-service teachers need to *experience* elements of classroom teaching practice, but this alone is also inadequate for comprehension as the mind is too caught up in the whirl of present activity to be able to rationalise what is occurring effectively and thus create useful mental models and generalisations on which to base future decisions. It is in the process of reflection after the practice, often best done with skilled support from a mentor, that the potential lies to obtain a new form of understanding about the principles which underlie what has been experienced.

The excitingly named American researcher and theorist King Beach (1999) describes learners as making a 'consequential transition' when engagement in learning experiences in one context bears fruits they are supposed to internalise and apply later in a new context. He criticises the notion of 'transfer' (of knowledge, skill or understanding) on a number of counts. Beach points out that learners themselves change in the course of learning experiences and that these changes cause changes in the nature of the tasks learners undertake and in the social settings in which they undertake them. So, for example, as the pre-service teacher becomes more skilled in designing learning activities for their own classroom, the mentor may occasionally borrow an idea from them, increasing the status of the pre-service teacher in the eyes of mentors and their colleagues. So there is causality in both directions – the learner is *transformed* through the learning experience; there is a process of *becoming*. This notion connects strongly with Wenger's notion of 'identity in practice' which he suggests is:

> a way of being in the world. It is not equivalent to a self-image; it is not, in its essence, discursive or reflective. . . . [It] is defined socially not merely because it is reified in a social discourse of the self and of social categories, but also because it is produced as a lived experience of participation in specific communities.
>
> (1998: 151)

Pre-service teachers, engaging with the practice of becoming teachers in the classroom, are involved in learning new ways of engaging with both other teachers and learners in the classroom. Pre-service teachers need to be accountable for their actions in this new practice and they need to negotiate a repertoire of actions that establish them as participants in the community of practice of teachers in the school in which they are placed. In this process their identities will follow a 'trajectory', a term which Wenger uses to describe a path of development that is continually changing and developing in ways that are not necessarily predictable or pre-determined (1998: 154). Becoming a teacher is seen as entering into a new community of practice and taking one's place in that community through the development of a personally distinctive set of knowledge, skills, beliefs and practices.

The attitudes of mentors towards supporting pre-service teachers

How many times has a colleague told you, in a slightly defensive tone perhaps, that the class you have inherited from them 'ought' to remember some or other mathematical idea because 'we covered that last year'? It is one of the most challenging demands of all upon any teacher to be brave enough genuinely to investigate the learning gained by people we have taught and accept the truth that our efforts have been largely in vain. The temptation is to blame the learners for their lack of effort or will. The fact is that we cannot transfer our own knowledge and understandings into the heads of others and our power and skill as educators lies in subtle and skilful plans and manipulations designed to help learners attend to the right matters, experience the right coincidences, notice the right patterns and so on. This notion of student-centred learning originates in the tradition of psychotherapy which sees therapy as a learning process (Rogers 1951: 132). Rogers elaborated a theory of personality and behaviour from his work as a psychotherapist which he then sought to apply to education. This application was widely drawn on in the development of notions of child-centred education in the 60s and 70s, as well as the development of theories of adult education or 'andragogy'. It is based on the premise that we cannot teach another person directly but only facilitate their learning. This also connects with Wenger's theories about 'learning communities' and his assertion that "*Learning cannot be designed:* it can only be designed *for* – that is facilitated or frustrated." (Wenger 1998: 229). Strongly linked to this is the notion that we cannot 'have all the answers'. A good mentor will need to embrace the fact that forming a pre-service teacher into a teacher like themselves by telling them how to do things is unlikely to be effective. Although the pre-service teacher may often try to adopt their mentors' teaching styles and language, in order both to gain their approval and as a way of becoming accepted as a teacher in the community of practice in which they are operating, this is unlikely to be successful at a deep level without support from the mentor to interrogate these adoptions and the beliefs that underlie them (Cain 2008).

Developing professional understanding

In order to become a full member of our community of practice, a pre-service teacher needs to be able to answer a host of '*why*' questions. It is in considering the why questions about their work that professionals become able to form the generalisations and rationales which constitute the body of their professional understanding: Why did you make that into a game? Why did you ask her as well when he had already answered? Why did you ignore it when she did that? As adult learners, pre-service teachers will want to explore these 'why' questions both through their own personal reflection and through discussion with their supporting mentor. The reality is that most of us find the answers to these 'why' questions difficult to articulate. This is often because we have been in the habit of making similar decisions in related situations for a long while and the understanding behind them has not been examined for some time. Equally likely, and perhaps scarier, is the possibility that we have picked up the habit from copying some role model in our past and that we have not really exposed it to the light of day to examine our rationale for it. To expose these assumptions and practices to scrutiny requires the mentor to develop his or her own capacity for critical reflection as well as encouraging the pre-service teacher to do so. The internalised critical voice that the reflective practitioner develops is only cultivated through a continual process of developing awareness and the capacity to notice the learning and behaviour of oneself and others in the classroom. Descriptors of these frames of mind and ways of thinking that accompany them have been clearly elaborated by both Caleb Gattegno (Gattegno, Brown & Tahta 1989) and John Mason (2002). In either case, in order to provide a useful learning conversation with a pre-service teacher we will need to be brave enough to dig deep, work outside of our comfort zone and struggle to find some words to express our own underlying rationales and ways of thinking. The decision that I will make myself struggle to elucidate for you my reasons does require courage, especially if I am a teacher who is proud of knowing answers in my classroom. The mentor needs to position him or herself in a certain way relative to the pre-service teacher and relative to the professional learning the mentor and mentee are engaged in.

Successful mentors *are* and *position themselves as* co-learners: there can potentially be a reciprocity in the mentor's engagement with their pre-service teacher in which each facilitates the professional development of the other. They are brave enough to admit to not having answers to all 'why' questions on the tip of their tongue. They are reflective in their own practice and model self-critical reflection for the pre-service teacher to witness. They have a 'learning' disposition as opposed to a 'fixed ability' disposition – which is to say that they believe in the potential for learners' 'abilities' to be transformed through learning experiences rather than sensing some fixed 'ability' (or even 'potential') in a person which they do not expect to shift. This notion of ability as malleable and open to cultivation through learning is expressed strongly by Carole Dweck's (2000) book on this subject. Her work refers to children as learners but is equally applicable to adults. She suggests that everyone can develop their intellectual abilities with effort and guidance.

In the context we are considering, of the pre-service teacher, this guidance will come in part from the mentor, with the effort to develop as a teacher being contributed by the pre-service teacher. The mentor needs to be brave enough (or comfortable enough about their own self-worth) to want to support a pre-service teacher in making the maximum progress in the brief time available – even when they have a sense that the pre-service teacher may become a more charismatic or better organised or more knowledgeable mathematics teacher than they are! Imagine the learners in your favourite class saying, "Oh, it's you, where's Miss today, she's great!" This is what you need to be prepared (potentially) to take 'on the chin'.

Good mentors expend the energy necessary to listen calmly and skilfully to the pre-service teacher, communicating a message of respect and the valuing of the mentee's thoughts. They ask gently probing questions, prompting the pre-service teacher to expand and reflect upon his or her ideas. They behave contingently in withdrawing support when they perceive that the pre-service teacher is capable of coping alone with some aspect of teaching. The goal in mind is always to support the pre-service teacher to become independent. So, assuming that you are willing to put yourself forward for such a challenge, we go on to look at some aspects of working with pre-service teachers as adult learners in schools before considering the practicalities of developing as a good mentor.

How adults learn

Traditional transmission views of learning and teaching are inappropriate in the context of adult learning. A traditional pedagogical view that sees the teacher as responsible for decisions about what is to be learnt and how it should be conveyed to the learner, as well as how learning should be assessed, is inappropriate in the context in which the learner brings to the situation significant life experiences as well as prior knowledge of which the mentor may only be vaguely aware. Although the mentor may have the responsibility for assessing whether the pre-service teacher has reached the required standard for success in the school placement, this responsibility is shared with others and can be negotiated with the pre-service teacher, who can be encouraged to develop personal strategies for formative assessment through guided reflections on his or her teaching practice.

In the context of the pre-service teacher on school placement there is a mutuality of engagement between the participants that implies a sharing of power and responsibility and echoes Wenger's theories about the ways in which novices and old timers interact in communities of practice (Wenger 1998: 157). Pre-service teachers need to be aware of the learning required to fulfil their new roles as teachers.

Pre-service teachers have their own sense of their identity into which they need to build their identity as a teacher. They bring with them a range of various experiences to the role, whether they are twenty-year-olds who have just graduated or mature entrants to teaching with another career behind them. As such, each pre-service teacher is likely to be very different from any other. In some cases the pre-service teacher may be older than the mentor and have held a position of considerable power and authority in another context; adjusting to their new role may be hard for both pre-service teacher and mentor.

The appropriate pedagogical approach, then, in working with adult learners such as pre-service teachers will need to be very different from a traditional transmission approach. One of the leading theorists in this area is Malcolm Knowles, who elaborated the concept of 'andragogy' as a theory of adult learning (Knowles, Holton and Swanson 2011). The assumptions that underlie his notion of 'andragogy' are that adults need to know why they need to know something; that adults have a strongly held concept of themselves and of being responsible for their own lives; that adults come to any new learning opportunity with a collection of prior experiences and knowledge; that adults are ready to learn when they see the relevance of the learning to their goals and intentions; that adults are motivated to learn when they see the learning as impinging on their goals and intentions; and that external motivation may be relevant to them when it is connected with these goals and intentions. In working with pre-service teachers it is essential that mentors take account of this independence and autonomy of the pre-service teachers and connect with them at an appropriately adult level. They cannot be patronised in ways which might be possible with younger students in school. Let us now consider the practical implications of these theoretical positions.

The importance of prior learning

When one of the authors asked a number of school-based mentors about the differences between their pre-service teachers and the children in their classrooms, they mentioned the pre-service teachers' prior knowledge of schools and schooling as significant. All prospective teachers have themselves attended schools, usually for at least 13 years, and have had ample time to develop tacit (wordless, unexplored) understandings of what mathematics is (see Chapter 3) and what teaching entails. If these understandings are left unexamined they have a profound effect upon the kind of teacher the pre-service teacher becomes. By their very nature (tacit), these understandings do not contribute to the development of a rationally defensible professional position unless they are scrutinised and subjected to interrogation. Pre-service teachers need to become more aware of their status as very successful learners in mathematics, whose success may well have been despite, rather than because of, some of the teaching approaches they experienced. Those with such prior experiences naturally tend to equate their success with the efficacy of the approaches they experienced. They need to develop the ability to move away from an ego-centric viewpoint which considers existing practice as 'good enough for me' and become increasingly attuned to the learning needs of the students whose learning becomes their responsibility. In most initial teacher education programmes, the pre-service teacher is prompted to consider and perhaps write about these beliefs very early on. The mentor is thus in a position to draw them out further through on-going discussions of issues. This process is especially effective if the mentor can quickly begin to establish an open and trusting relationship with the pre-service teacher in which the pre-service teacher feels comfortable to discuss ideas honestly and in depth.

Peer support for pre-service teachers

There are several other sources of support pre-service teachers need to be able to draw on whilst on school placement, including the need to discuss experiences with fellow pre-service teachers, where possible. These conversations, focused as they are on the same set of problems being experienced by two or more pre-service teachers at once, can often be rich sources of learning and contain none of the sense of hierarchy amongst participants which we have already implied can be unhelpful. This is elaborated further in Chapter 6. A school may be hosting more than one mathematics pre-service teacher, or the only other pre-service teachers in the school may belong to other departments or teams within the school – in either case mentors can facilitate some of these learning conversations by ensuring that the pre-service teachers have some shared 'free time' and perhaps also by prompting them to discuss with each other a particular issue at a particular time. Engagement by pre-service teachers with peers through participation in online learning forums can also be valuable in developing awareness of professional identity.

The mentor and pre-service teacher relationship

Who do you feel at ease with when discussing ideas related to your work? People who are friendly towards you and whom you trust not to have any agenda which might not be directed in your best interests? People who are not prickly and do not try to outscore you or belittle you with displays of their superior knowledge? People who know a little about you as a person outside of the professional sphere, but who are not so emotionally close to you that discussion might become stilted through carefulness of their feelings? To set up such a relationship fairly swiftly requires of the mentor both the dispositions we have already looked at and a significant amount of time and effort. Have you prepared for the arrival of your mentee(s), creating a draft programme of observations for them so they can experience a range of teacher styles and learner attainment? Have you thought about what they will do at various times during their first day in the department, year team or school? Will they get a cup of tea and a biscuit? Somewhere to sit and to work? An introduction to the members of the mathematics department or year team? Will there be a chance to speak to you at some length before the end of the day so that they may ask some of their most pressing questions and go home with a sense of what tomorrow holds? Will you provide some introductory guide(s) to your school and its policies? Will they be invited to attend meetings in school, and any upcoming social event outside school? There are other books which offer detailed suggestions for how mentors can plan programmes for pre-service teachers (see, for example, Fletcher 2000); what we are more concerned with here is identifying the *types* of practical steps that mentors can take to help the mentee quickly to feel at home, a valued member of the team and trustful of mentor support. Cain (2008) draws attention to the emotional tension that pre-service teachers experience as they try to fit in to their school community whilst at the same time being true to themselves and their ideal of the kind of teacher they want to be. His findings suggest that the mentor needs to work hard on developing their sensitivity to the pre-service teachers' concerns and conflicts in all their dealings with them.

One important forum for learning conversations is the weekly meeting between pre-service teacher and mentor. The pre-service teacher should be encouraged from early in the placement to bring to this meeting an agenda of points for discussion along with an assessment of current strengths and thoughts about areas for improvement. In paired placements the inclusion of both pre-service teachers in these meetings can powerfully add to conversations. The pre-service teacher could be using some form of 'reflective journal' to record their ideas and develop understandings about aspects of teaching. If excerpts from this can be discussed as part of a weekly meeting its power as a working document can be augmented. If the journal is completed in an online format, the potential exists for regular inclusion of the supervising tutor who may receive email updates of additions to the journal – creating a potential 'third space' if the tutor can also contribute ideas.

Supervising tutor–mentor–pre-service teacher relationship

In many forms of pre-service teacher education an accrediting institution also provides a supervising tutor. Maintaining contact with this tutor amongst the busyness of life in school can be challenging and we examine briefly a few ideas as to how this can be facilitated. Before doing so, however, it is important to emphasise how the nature of the tutor–mentor–pre-service teacher relationship can influence learning. If a learner has a strong sense of hierarchical positioning relative to another in a conversation they may well be influenced to modify what they say. Not wishing to appear ignorant or ill-informed they may stifle questions which occur to them. They may try to take on board points which they feel they have not understood well, out of some deference to their originator. All of this is antagonistic to learning. Our goal is for the pre-service teacher to become a self-governing reflective practitioner, capable of managing their own on-going professional development as they move on through their career. They need to move towards a position where it is they who prioritise their own learning needs and take the lead in setting agendas for mentoring conversations. Recently, research has started to examine opportunities for the creation of 'third spaces' (Gutierrez 2008) – arenas in which 'horizontal' or non-hierarchical conversations may be facilitated between these three parties. In these conversations, each participant is positioned as a co-learner, each feels able to voice their thoughts without fear or favour. Another potentially very powerful third space can be created through video analysis of a pre-service teacher's teaching.

The use of video recording to provide a means and a set of opportunities for reflective practice is increasingly well documented. In particular, the work of Rachel Lofthouse and Peter Birmingham (2010) describes how watching video of lessons led by the pre-service teacher can help to provoke collaborative professional dialogue as it focuses the participants' attention on the detail of events that actually happened and gives substance to their analysis of the processes of teaching and learning as they can be observed. The pre-service teachers themselves commented that they found the evidence of the video helpful as they could see what they were doing and understand their mentor's comments more easily. In addition, Youens *et al.* (2012) describes how this can be taken on to a new level via asynchronous review of video in a forum chaired by the pre-service teacher. In this model, video of one or more of the pre-service teacher's lessons is recorded. They are required to review the lesson, first drawing only upon their own recollections, then with the aid of the video. They then select excerpts of the captured video and use them as the basis for a three-way learning conversation (or four-way if a fellow pre-service teacher is invited) in which they direct the attention of mentor and supervising tutor to particular aspects of lessons, focusing on priorities of their choice. This shift of responsibility to the pre-service teacher facilitates more of a 'level playing field' in the conversation. It also makes it more likely that the pre-service teacher (perhaps with some prior prompting from mentor or supervising tutor) can focus on underlying theories or beliefs about school students' learning which led them to make particular pedagogic decisions. The nature of the 'evidence' here and the fact that the action may be paused or re-run is seen as assisting in this depth of analysis. In many cases the success of this practice can lead to it being employed by other members of a department or year team to further their own professional learning.

The fact that teaching is such a complex endeavour means that to learn how to teach effectively from a reasoned position takes a very long time. The mentor's awareness of the very many steps involved and the many competing distractions which can serve to make the process more difficult or the learning less effective is central to supporting the pre-service teacher.

The complexities of learning to teach

When a person first learns to drive, the range and number of things to think about all at once is usually completely overwhelming. Gear changes can be accompanied by dangerous swerves, signalling by a lack of attention to the road ahead and so on. Over time, the learner assimilates more and more of the elements of car control and finally becomes able to drive whilst paying effective attention to the unfolding scenario around them. Teaching is similar in that the number of disparate elements to be attended to is huge, more than enough to overwhelm the brain's capacity for concurrent awareness in the early stages. As we develop in our professional understanding and skill, we internalise more and more of the regular functions and practices and become capable of applying a greater degree of awareness to the unfolding needs of our learners. The work of one of the authors using lesson study approaches offers an illustration of this principle (Hunter & Back 2011) and there are resonances between this approach to observing and analysing lessons that can be replicated in the mentoring setting. Mentors can use this understanding in order to challenge pre-service teachers in incremental stages. Perhaps they might observe some lessons at first, reflecting upon how little of what they see they can actually comprehend in terms of what skilled behaviour on the part of the teacher led to elements of the lesson appearing as they did.

Critical engagement of the pre-service teacher and the mentor in analysing the planning process also connects with the kind of in-depth analysis that a lesson study approach uses (Hunter & Back 2011) and can support the pre-service teacher in continuing their professional development further as they continue teaching. Later they might plan to teach a small section of a lesson. The planning is likely to take them a long time – the planning prompt sheet used by one of the authors' higher education institution includes some 80 individual prompts for aspects the pre-service teacher should consider at the planning stage. Having taught this episode and reflected upon it with the mentor, the pre-service teacher is in a position to notice more effectively what the teacher of the next lesson they observe has done at the planning stage and live in the lesson to bring about the outcomes witnessed. This cycle will continue. The mentor gauges the pre-service teacher's readiness for next steps, keeping the number of targets to be worked on at any given time to a manageable number (three or so), reviewing them alongside the pre-service teacher and supporting them in setting more targets. It is important that these targets are accompanied by the actions needed to move towards them – pre-service teachers can be bewildered by suggestions that they need to 'increase their classroom presence', for example, but assisted by suggestions as to how this may be achieved – 'by positioning yourself in these three places in the room at these points in the lesson'.

Affordances and constraints

Mentors should be aware of the tendency to remain inside their own comfort zones. Teaching effectively is undoubtedly strongly influenced by the relationships that a teacher has established with the teaching group. Since the mentor always tends to be comfortably established in these relationships, at least relative to the pre-service teacher, it is tempting to over-emphasise them as a (subconscious) means of maintaining a position of comfortable superiority. The same can be applied to classroom 'confidence', another area in which the mentor is always comfortably ahead. The pre-service teacher needs to arrive at a position where they can move on to a new teaching post with some degree of confidence, taking their professional learning with them. This knowledge and understanding equips them to practise in a new environment basing their practice on making rational professional decisions.

A common issue is that of mentors failing to notice the ways in which pre-service teachers have made progress since they began. Like all learners, pre-service teachers need to receive positive strokes to help maintain their motivation and it is all too easy for a mentor to overemphasise criticism through the best of motivations. Having pre-service teachers first review their own thoughts about how a lesson has gone, and tempering what observations you do offer with noticing of desired learning outcomes help the pre-service teacher to remain resilient and keep striving to improve.

Personalised approaches to mentor support

Many pre-service teachers are 21 or 22 year olds who have just graduated from their undergraduate degrees. Readers may be able to cast their minds back to their own undergraduate years and consider the potential mismatch between an undergraduate lifestyle and that of a successful pre-service teacher. Mentors may well need to support the young pre-service teacher in understanding and adapting to the demands of a professional working life. Organisation of the pre-service teacher's own time (including the need for healthy exercise, rest and some relaxation), prioritisation of tasks and development of helpful routines and effective organisation of resources are all aspects which may not come quickly or naturally without some explicit coaching and modelling from the mentor.

The fact that this group of pre-service teachers may be relatively close in age to the learners they teach can present issues of its own. They may well have interests in common and feel at ease early on in holding conversations with them about these. This may combine unhelpfully with the natural desire to be liked that we all feel and a possible lack of maturity in handling conflict. If not closely supported, the pre-service teacher may thus fail to create appropriate 'professional distance' at the outset and be particularly ill at ease when needing to enforce classroom consequences – which, as we know, can have serious consequences of its own for classroom ethos. Some may also be guilty of not fully thinking through their actions and need help in identifying the results of some hasty response and in acquiring more careful habits of mind for the future.

Developing effective emotional intelligence may well be a particular priority for some young pre-service teachers. They need to become aware of the potential for a low mood, resulting from some critical feedback or a lesson which did not go to plan to affect their performance in subsequent activities in that day. This requires great gentleness and sensitivity of the mentor.

Pre-service teachers who are mature adults, career changers and perhaps parents bring a different set of perspectives, needs and external responsibilities to those who have come in to pre-service directly from completing their own education. It is worth considering who mentors each pre-service teacher as perhaps a match may be found which is likely to result in better appreciation of the other's issues. Mature career changers may have come from a successful career in another field in which they have become used to the status and respect accorded to an expert. They may thus find it exceptionally challenging to return to the status of novice in the field of teaching. This may involve mentors in some discussions around areas of expertise and the possibility of being a novice in one field and an expert in another – perhaps the pre-service teacher has some expertise which their mentor might learn from or show interest in. To seek guidance from someone with less life experience than themselves may be difficult and they may recognise (see through) some of the strategies mentors employ in target setting and other forms of support. It may be best to broach these topics openly at the outset as channels of communication about them will need to remain open as the placement progresses. These learners may also be used to analysing tasks in great depth in a work context and may thus become over-immersed in some or other research task set as part of their learning programme. University leavers may be more adept at cutting corners sometimes to get tasks finished by deadlines – which can be both a strength and a weakness.

Another issue for the mentor may be conflicts between the goals of the school in which they work and the process of supporting the development of new pre-service teachers. The mentor may need to ensure that the attainment scores of the learners they teach are as high as possible, whilst pre-service teachers will rarely have entered the profession with this as one of their priorities but are more likely to be concerned with learners' enjoyment or understanding. Recognition by the mentor that they too are aware of this conflict in priorities is important. Hopefully they may be able to help the pre-service teacher to discover a pedagogy which is engaging and results in lasting understanding of mathematical ideas as well as helping learners to succeed in tests. If not, then acknowledging that this may be possible at some point in the future may help all to maintain their sense of professional identity.

Developing the placement experience

As the placement continues and the pre-service teacher's skill level increases, the mentor is able to focus observations of lessons increasingly upon particular aspects of practice agreed at a previous meeting and prioritised by the pre-service teacher. Post-lesson discussions then contain a greater proportion of talk by the pre-service teacher as the mentor hands over more and more of the responsibility for reflection to him or her.

The mentor may resume teaching the class for a lesson or two at any point in order to model a teaching method which has been highlighted in discussion, or they might suggest that the pre-service teacher takes the opportunity to observe a colleague in the same or another department or year team who is particularly highly skilled in this area.

The mentor will be aware that the pre-service teacher learns some lessons far more effectively by making mistakes, so does not always seek to have amended every flaw they detect in a lesson plan. By subsequently supporting the pre-service teacher in reflection about causes and effects they can facilitate deeper and more rapid learning. Creative ideas which the pre-service teacher wishes to try out are encouraged – this is the time when experimentation can be most freely engaged in, as there is a supportive professional in attendance to help out if things should go awry. A pre-service teacher who is engaged in effective professional learning will make plenty of mistakes. These may sometimes lead to crises of confidence and the mentor may be called on to be emotionally supportive of the pre-service teacher at various points along the path, a steadying presence as they experience the rollercoaster ride which learning to teach almost always resembles.

Final remarks

In this chapter we have identified some of the models for learning which can provide a framework for our work with pre-service teachers. Pre-service teachers are seen as learning to become a teacher through reflective practice, making a consequential transition from peripheral participation in our community of practice to a more central role in which they have a personal professional identity. The learning, we have argued, is maximised through the mentor being positioned as a co-learner, exposing prior knowledge and beliefs, joining in the reflective process and using their accumulated wisdom to help choose or design experiences predicted to help the pre-service teacher to take a next step. We have advocated cooperative non-hierarchical relationships between pre-service teachers, mentors and supervising tutors and identified spaces for learning conversations between the three. Finally we have examined a number of issues particular to working with pre-service teachers in schools, and some possible responses to them.

To be a mentor, creating supporting structures for the pre-service teacher's learning experience, working alongside them contingently, welcoming them into the school, department or year team and profession with a view to helping them to be the best teacher they can be, is not only an extraordinary challenge, but one which will have profound consequences into the future in influencing many thousands of learners' lives for the better.

Acknowledgements

Mark Simmons would like to acknowledge the contribution of colleagues at the School of Education at the University of Nottingham, especially Professor Peter Gates, and mentors from its partnership schools, particularly the Fernwood Academy.

References

Beach, K. (1999) 'Consequential transitions: A sociocultural expedition beyond transfer in education', *Review of Research in Education*, 24, 101–39.

Cain, T. (2008) 'What can research tell us about mentoring trainee teachers?' in D. Fischer, L. van Andel, T. Cain, B. Zarkovic-Adlesic, & J. van Lakerveld (eds.), *Improving School-based Mentoring: A Handbook for Mentor-Trainers*, Munster, Germany: Waxmann, pp. 25–38.

Dweck, C. (2000) *Self-Theories: Their Role in Motivation, Personality, and Development*, New York: Psychology Press.

Fletcher, S. (2000) *Mentoring in Schools: A Handbook of Good Practice*, London: Kogan Page.

Gattegno, C., Brown, L. & Tahta, D. (1989) *A Gattegno Anthology: Selected Articles by Caleb Gettegno*, Reprinted from Mathematics Teaching, Derby: Association of Teachers of Mathematics.

Gutierrez, K. (2008) 'Developing a sociocritical literacy in the third space', *Reading Research Quarterly*, 43(2), 148–64.

Hunter, J. & Back, J. (2011) 'Facilitating sustainable professional development through lesson study', *Mathematics Teacher Education and Development*, 13(1), 94–114.

Knowles, M., Holton III, E. & Swanson, R. (2011) *The Adult Learner: The Definitive Classic in Adult Education and Human Resource Development*, 7th edn, Oxford: Butterworth-Heinemann.

Lave, J. & Wenger, E. (1991) *Situated Learning: Legitimate Peripheral Participation*, Cambridge: Cambridge University Press.

Lofthouse, R. & Birmingham, P. (2010) 'The camera in the classroom: video-recording as a tool for professional development of student teachers', *Teacher Education Advancement Network Journal* 1(2). Online. Available HTTP: < http://bit.ly/tyfJ5M> (accessed 1 January 2013).

Mason, J. (2002) *Researching Your Own Practice: The Discipline of Noticing*, London: Routledge Falmer.

Rogers, Carl R. (1951) *Client-Centered Therapy: Its Current Practice, Implications, and Theory*, Boston: Houghton Mifflin.

Schön, D. (1983) *The Reflective Practitioner: How Professionals Think in Action*, Washington: Basic Books.

Wenger, E. (1998) *Communities of Practice: Learning, Meaning and Identity*, Cambridge: Cambridge University Press.

Youens, B., Smethem, L., & Sullivan, S. (2012) *Promoting Collaborative Practice and Reciprocity in Initial Teacher Education: Realising a Dialogic Space through Video Capture Analysis*. Paper presented to the *Universities Council for the Education of Teachers* annual conference, Hinckley, 8–9 November.

Chapter 11

Turning the mathematics learning environment into a research environment

Alf Coles and Laurinda Brown

We introduce the chapter with reflections on what we mean by learning and what we mean by research: for us, research is a form of learning. We describe how we have each set up structures that support others in developing as teachers, through a process where the teachers are supported in generalising, having first paid close attention to the detail of their experience, a process we see as researching practice. We explore aspects of how practice-based mentors have engaged in the journey of researching their own practice and how this engagement, in turn, has supported them in mentoring others, leading to the creation of a wider research environment in their institutions. We see the difference between a learning environment and a research environment as marked by whether there is an awareness of the learning that is taking place. A striking finding is that mentors who themselves have learnt to teach by researching their own practice display such awareness by supporting the research journeys of others, rather than offering models of practice to follow.

Introduction

This chapter is about learning and researching and the creation of particular environments; environments for ourselves, mentors, in-service teachers, pre-service teachers and students; environments in which to learn mathematics and about teaching mathematics. The chapter is structured into three sections: we first set out and illustrate what we mean by learning and researching; we then analyse the results of interviews with five mentors who work with us and with pre-service teachers and who have been involved in research; finally, we return to the question, what does it mean to turn the learning environment into a research environment? We draw on our experiences as tutors on a one-year Post Graduate Certificate of Education (PGCE) course in secondary mathematics as well as our work, in many different contexts, facilitating groups of in-service teachers. We both came to work at the university after fifteen or more years teaching in secondary schools.

What is learning? What is researching?

We begin with a story.

A mixed attaining class of 11- and 12-year-olds were in the middle of an investigation related to perimeter and area. Alf was their teacher. He had started the project by giving students the problem of finding the rectangle with the largest area, given a perimeter of 12cm. Having solved this initial problem, Alf encouraged students to try other perimeters, begin looking for patterns and start generating and working on their own questions. Laurinda came to visit the class. At the start of the lesson we invited students to share what they had found out so far. We got the following responses:

- The largest area for any different perimeter is a square.
- To draw a rectangle with a perimeter of an odd number you must use halves.
- Odd sides means an odd area.
- Even sides implies an even area.
- Divide the perimeter by 4, then times by itself, what you get is biggest area.

These statements were written on the board as they were said and no explanations were asked for or given. When the list was completed, Laurinda added that she had been thinking about the first question and wondered whether it was possible to find the perimeter of a square when the area was 50cm^2. There was an invitation to stay with what they were working on or incorporate any of these ideas into their investigations. The class continued to work individually or in small groups.

Sarah's story (from Brown & Coles 2008: 46–7)

After some time Sarah, who had been working on the 50cm^2 problem, came up to Laurinda stating that the perimeter must be 4cm. Laurinda drew a square with area 1cm^2 on the board with 1s marked around the perimeter (see Figure 11.1) and waited for some response.

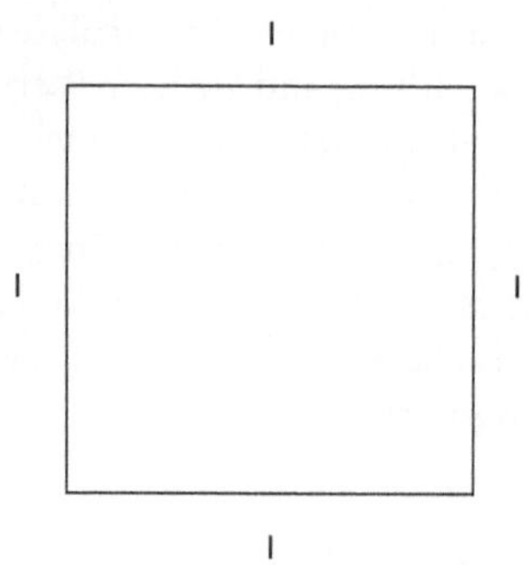

Figure 11.1 The sketch of the square shown to Sarah

When none came, Laurinda asked Sarah how she had worked the 4cm out. Sarah talked about 'reversing the rule *divide the perimeter by 4 then times by itself to get the biggest area*'. Laurinda started writing the flow diagram for this as she spoke and Sarah reversed 'times by itself' to 'divide into itself' and 'divide by 4' to 'times by 4'. This gave 50 / 50 × 4 = 4.

Laurinda said, 'OK, times by itself' and wrote: 4 →

Sarah quickly replied '16' having had experience of writing functions in this way, and Laurinda wrote 16 next to the arrow, followed by: 6 →

The answer '36' again came quickly. Laurinda offered: ← 49

Sarah responded with 7 immediately. They agreed that this had not been 'divide into itself' but what was it? Sarah went off to work on this.

We offer this story in order to illustrate how we think about both learning and researching. In Alf's classroom environment, students were used to sharing their ideas and seeing these as conjectures for other people to work on. The students were also used to asking their own questions and finding ways of structuring their mathematics. They expected mathematics to mean something, and new skills, new ways of thinking were occasioned through what could be called researching as a team.

There is a difference between how Sarah articulated the inverse of 'times by itself' ('divide into itself') and how she acted when faced with reversing a 'times by itself' arrow (she accurately gave the square root of 49), in the context of what we call the function game. She is faced with a difference. Sarah apparently recognises that her earlier 'divide into itself' is not the inverse function she actually performed. Whenever we see someone (ourselves included) responding differently within the same context, we see learning. Sarah at first saw 'divided into itself' as the inverse of squaring and, at the end of the story, does not. We do not assume anything about what is going on inside her head, nor that changes we observe in this story will mean she responds in the same way next time she is faced with a similar question. However, in this story, there is evidence of change and hence learning and also evidence of exploration that we see as research. Sarah was working on a difference in her mathematical ideas and Laurinda was working on how students learn mathematics and her own development as a teacher.

We first wrote about this story in 1997 (Brown & Coles 1997), at the start of what has been an on-going research collaboration. We were struck by Laurinda's responses to Sarah and what we could say about what was motivating Laurinda's actions. The paper we wrote was accepted for the 21st Proceedings of The International Group for the Psychology of Mathematics Education (PME) Conference and was therefore recognized as research. Our work seeks to focus on the details of practice, hence the way we try to tell the story with minimal explanation and interpretation. We discussed the incident at the end of the lesson in which it took place, given that Laurinda had been surprised at the immediacy and effectiveness of her own response. Laurinda had been aware of a difference between her image of the inverse of squaring and what Sarah articulated. She presented Sarah with images of that difference, initially unsuccessfully through drawing the image of a square that was ignored by Sarah and, then, through starting with a function that Sarah was comfortable with, followed by reversing the direction of an arrow. In this story, Sarah changed her initial response (the inverse of squaring was seemingly no longer seen as divide into itself).

For us, researching is about working on differences, staying with the detail of our experiences and trying to say something about those experiences. Our aim, in researching, is to inform our actions as teachers and researchers. In other words, we see research as a process of developing awarenesses, of seeing more and of opening up new possibilities for action. Hence, for us, research is a form of learning. We look for pattern, and for pattern in those patterns (see Bateson 1979: 10). We engage in research in order to change and we tend to write up our work emphasising the journey, making use of story (for example, see Brown & Coles 1997, 2000, 2008). As part of this research journey we explore different ways of seeing the same thing (Laurinda did not have only one way of offering Sarah a contradiction to her thinking) and we encourage students in classrooms to see mathematical ideas in different ways.

This image of teaching mathematics within a research culture also presupposes teachers who can adapt to what their students bring them from the range of strategies they have at their disposal. The teachers are learning about their students learning mathematics. The link between theory and practice, research and action is strong. For the students in the classrooms learning mathematics, algorithms are not presented separately from the underlying structures that give meaning to the ideas. The students expect to work on meaning themselves, rather than accept what the adults tell them uncritically. Similarly, pre-service teachers on our initial teacher education programme do not expect to be told by us what to do in their classrooms, but to explore the implications of what they do, for the learning of their students. In the story, Laurinda's mindset was to present Sarah with things to do or engage with, not to tell her she was wrong. For us, both in working with students and pre-service teachers, we believe that people act out of their previous experiences of the world, so, if something is not working, offering learners something to do that sets up a difference supports their learning.

We take these beliefs about learning and researching into our work with teachers and in the next two sub-sections illustrate how we, in different contexts, have set up structures to support others in developing as teachers through researching their own practice.

Laurinda: small group discussion on a PGCE programme

Over a series of papers and book chapters (Brown & Dobson 1996; Brown (with Coles) 1997; Brown & Coles 2000; Brown 2005) Laurinda articulated her emerging practice as a teacher educator. She observed that, in discussions with her students who were pre-service (beginning) teachers:

> Beginning teachers need to temper idealistic goals given the reality of how much skill might be required to achieve them . . . engaging with a student on

> a philosophical level . . . did not seem to allow practical development or change of implicit theories or attitudes (Claxton 1996); nor did giving 'tips for teachers' at a behavioural level do much for their developing sense of who they might be becoming as a teacher.
>
> (Brown (with Coles) 1997: 104).

What did seem to be effective was guiding discussion not in terms of content, but in terms of form, or structure, to a place between the abstract and the detail. Laurinda wrote how she

> started to use the word 'purposes' as an emergent description of the sorts of guiding principles that student teachers found energising when learning from their own experience. Chamber's Twentieth Century dictionary (1976) defines purpose as: *idea or aim kept before the mind as the end of effort; power of seeking the end desired*. There is nothing here about actually getting to the end, although that is not precluded. With purpose, however, we give ourselves the motivation to make effort in relation to some 'idea kept before the mind'. Purposes seemed to be in the middle position between philosophical attitudes and teaching behaviours in the classroom.
>
> (Brown 2005: 1)

The Mathematics PGCE course at the University of Bristol was set up (by Laurinda) to help pre-service teachers identify and then work on purposes in their own practice, and to extend their range of practices beyond those which come naturally. The mathematics course documentation begins with: 'There is not one way to teach mathematics' and one statement from the Master's level assessment criteria for the whole PGCE course, reads 'assurance in handling a range of perspectives'. Whether we are doing mathematics together or sharing teaching strategies such as questioning, the pre-service teachers are in groups and sharing their practice, just like the students in Sarah's story. We deliberately give PGCE places to individuals from the widest possible range of backgrounds and ages, from engineers to accountants, from doctors of pure mathematics to mechanics specialists, from tree surgeons to teaching assistants, all with success in mathematics before they come on the course. Instead of judging another's contribution (as a stereotype, engineers are comfortable with estimating and pure mathematicians can like to 'be right') we encourage our pre-service teachers to imagine themselves being different, because students do not learn mathematics in the same way. Whilst any teacher's natural way of thinking and explaining might work for some students, it will not work for all.

From the start of the course, our expectation is that the pre-service teachers act as 'teachers'. Our experiences tell us that novices and experts can operate in the same way (Brown & Coles 2011). The course begins with two weeks of observations in a primary school and we work with those experiences as the pre-service teachers arrive at the university. Whenever they return to university from a period of time in school, either we get them reflecting in a tutor group (each of us has about 15 students whom we work with on school placements) or they sit in threes as a reflecting team, talking through their teaching experiences. In the tutor groups, after hearing the first story from experience, the invitation is then for another pre-service teacher to offer a story that is sparked off as either being the same in some way or contrasting. The collection of stories continues until a common theme has emerged. The university tutor at that point might suggest a focus on 'What are we talking about?'. The idea of this phase of discussion is to arrive at the articulation of a purpose, such as: 'how do I know what the students know?', or 'how do I deal with emotions?', or 'how do I involve everyone?'. Having arrived at a purpose there is then a sharing of as wide a range of teaching strategies as possible, linked to the purpose. We insist pre-service teachers raise to awareness what aspect of their teaching they are working on, at any stage on the course. We encourage the articulations of these targets to be in the form of purposes. These purposes can then be 'kept before the mind' to support both action in the classroom and reflecting afterwards.

Alf: working with video

From 2001 to 2008, Alf worked as the head of mathematics in a mixed 11–18 secondary school in the UK. After his involvement in a research project (see Brown, Sutherland, Winter & Coles 2001) that made use of video recordings of lessons, Alf decided to continue the practice of taking video recordings within the department and also to dedicate some teacher meeting time to watching excerpts of these videos of each other's classrooms.

There is a common problem in the literature on using video with teachers, which is that discussion often takes the form of evaluation or judgement, resulting in little of value arising for the teachers involved (see, for example, Jaworski 1990; Nemirovsky, Dimattia, Ribeiro & Lara-Meloy 2005; van Es & Sherin 2008). In order to overcome this problem, a particular practice was developed at the Open University (OU) in the 1980s (Jaworski 1990) where, having first watched a small clip of video (3 minutes long), the initial task for the viewing teachers is to try to reconstruct, in as much detail as possible, the events that took place. The idea is to become engaged in the fine detail of what was said or done. The facilitator of the discussion may have to intervene, initially, to keep the talk focused on this detail as it is easy to slip into judgements and interpretations of what was seen. After time spent in the detail of these events, possibly re-watching the clip several times, there can then be a move in the discussion to considering interpretations of events.

If discussion of video moves straight away into interpretation, participants can only talk in terms of distinctions they already use. Jaworski (1990) gives the following as examples:

> 'He was railroading them – they didn't have a chance to think for themselves.'

> 'I couldn't do that with my pupils, they can't work quietly enough;' 'they're not intelligent enough;' 'they don't wear uniform;' 'we can't arrange the classroom like that;' etc.
>
> 'I could never do that – I just don't have the right sort of personality.'
> (1990: 63)

As Jaworski went on to analyse, the problem with such comments as the ones above is that they are interpretations or judgements that seem to be based more on viewers' own experiences than the events on the video. Jaworski (1990), importantly, noted that if such comments were made by the first speaker in discussion of a clip, then those comments can influence the rest of the discussion and result in little of value taking place for the participants. Yet, these kinds of interpretation are how we generally speak and notice – we use familiar categories and see things in relation to previously held ideas and beliefs. But if we begin discussion by using categories and concepts that are familiar to us, it is unlikely that we will notice anything new; what we see will conform to our previous ways of thinking. As humans, we are very good at making the world fit the way we think it should be!

In contrast, when discussion begins with a period of time spent reconstructing events on a video, there is a greater possibility for the making of *new* distinctions. It is important to move into interpretation of the video, but if this has been preceded by time spent on reconstruction, the analysis has a good chance of throwing up new categories or ways of seeing (see Coles 2012 for further discussion of the role of the discussion facilitator in working with video). In other words, the OU model of using video supports teachers in learning. Alf adopted this OU model of using video in his work with teachers in the mathematics department (Coles 2013). The aim, in moving to interpretation of the video was to arrive at potential 'purposes' that teachers could then work on in developing their own practice.

Similarities and differences

It is unsurprising, given we have collaborated since 1995, that there are strong parallels between our work in the contexts above, on a PGCE course and working with video. One strong similarity is our conviction about the need to begin discussions with an articulation of the detail of experience and then moving to generalisation or identification of a purpose. The focus of the university tutor on the PGCE course, or the facilitator of discussion of video, is firmly on the noticing of the 'other'. The tutor or facilitator needs to be alert to when stories or comments drift into interpretation or judgement at the start of a discussion, or, later on when someone articulates something that could be a purpose. We see this stance of the tutor or facilitator as central to the learning of the teachers with whom we work. With our attention on the learning of the teachers, they are supported in becoming aware of differences between how they acted and how they might have acted. The awareness of different possibilities for action being linked to the articulation of a purpose also supports teachers in trying out new actions. We see this cycling of an awareness of a discomfort or distinction about a past action, leading to an articulation and then a different way of acting in the future, as one of the hallmarks of research.

Stories of practice from school-based mentors

Having set out our own views on learning and researching and illustrated how, given these views, we work to support the research journeys of others, we now turn to examine the work of other mentors. We work with school-based mentors in our roles as tutors on the University of Bristol PGCE and Masters' courses and, in some cases, in the context of collaboration on a research project. We contacted mentors who had been recently involved with us in research to ask if they would take part in our own research for this chapter. Laurinda audio recorded interviews with five teachers to investigate their approach to mentoring and researching. Before any recording was done, Laurinda discussed the nature of the interview and the fact that excerpts would be used in this chapter. All mentors (identified by pseudonyms) were happy to take part. We devised a core set of three questions (what's research for you?; how does this fit with you work in the department?; and how does this research impact on your mentoring of PGCE pre-service teachers?) to ask for views about research and mentoring, but once the interview began, Laurinda engaged in a dialogue to probe the mentors' responses. In this section, we report on the results of these interviews. Having transcribed them all we became aware of some strong similarities in responses (as well as some differences). As we stated earlier, we see our research in terms of an investigation of difference and pattern, hence these ideas inform our data analysis. We report here on two patterns in the responses which we have labelled: 'research as developing teaching practice' and 'reflecting with pre-service teachers'.

Research as developing teaching practice

We asked the mentors what 'research' meant to them. In all cases, they spoke about research in terms of developing their own practice:

> To improve my practice so that I can improve the learning of the students. (Helen)

> I'm trying to develop myself and maybe that'll help other people develop. If the things I'm wanting to say are interesting to other people that's great but . . . it feels much more personal somehow. (John)

> I think for me it's been something that I've wanted to do to try and tackle an issue or just improve my teaching. It's been something I've done with a goal in mind, something that I want to improve. (Amy)

> I guess the stuff that matters from a selfish point of view is the stuff that develops my practice and less about what I report back on. (Adam)

> For me it's always been about developing my practice; looking and reflecting on my own practice in order to change something I want to change. (Sasha)

All the interviewees had been through the University of Bristol PGCE course and all had done a Master's in Education at the University of Bristol with either Laurinda or Alf as their dissertation supervisor. We see a striking similarity of views in the expressions above. In all cases, the mentors see researching as becoming conscious of an aspect of their practice they want to develop and then focusing on that. One mentor made a connection that this is precisely what he was encouraged to do on the PGCE course:

> I guess I see part of research is what I take from it. The [Master's] dissertation I took a lot from and I see parallels with that and working with other people to develop what I do was exactly what I did on the PGCE course . . . Of all the experiences we've talked about the things I guess the stuff that matters from a selfish point of view is the stuff that develops my practice and less about what I report back on. I don't know if that makes it research but that puts the dissertation and the PGCE [course] in a similar category for me. It's easier to label the dissertation as research. There are a lot of similarities in the way I worked for that and the way I worked in the PGCE year. [The PGCE course] is probably more applying what you gained in the classroom and more direct and quicker because you didn't have a base on which to try things, everything was trying things, but the working together with other people and then having those reflections challenged and then taking those reflections was part of the PGCE course. Part of me wonders whether the work on the Master's units and developing that way of working now makes me see that more than I would have done. If you'd asked me if [the PGCE course] was research before my Masters I probably wouldn't have thought of that as research then. (Adam)

We are struck by Adam linking his work on the PGCE course to doing research. Given our views on what it is to research, we would also see the course as a research environment, but would never articulate this to the pre-service teachers – we are not wanting to provoke in them any awareness about the course set-up that may get in the way of their participation on the course. What we do articulate to the pre-service teachers is that the year is about them 'becoming the teacher they can be'. The pre-service teachers need to become absorbed in their own practice and its development, not the fact that this is what they are doing. Adam only makes the link that the PGCE course was actually research as well after having done a Master's where he recognises that the processes are essentially the same: 'working with others', sharing reflections, 'having those reflections challenged', 'applying what you gained in the classroom'.

Reporting on research to others is mentioned by two mentors (John and Adam) but it is almost as though anything anyone *else* takes from research is an unexpected bonus; the work is done for the individual. We are reminded by this sentiment of Mason (2002) discussing how perhaps the most important product of research is the change produced in the researcher. He coined the phrase 'researching on the inside':

> [w]hat is important in qualitative research in general, and in researching from the inside in particular, is not the validity or accuracy of the description, but the effect, the action that the description sets up inside others.
>
> (2002: 229)

We see the outcome of research not in connection to other people doing the same as we do. Research will have an impact on others if it sets up new distinctions, new ways of seeing and hence new actions. Adam is aware that on both the PGCE course and in his Master's studies he was supported in acting differently, for example by having his reflections challenged. The link to action we see as vital: Adam talks about 'applying what you gained' and 'trying things' in his classroom. For us, and evidently for the mentors we interviewed, research is not research if it does not lead to *doing* something differently.

Reflecting with pre-service teachers

We asked the mentors to describe how their involvement in research has affected their work with pre-service teachers. There was a strong pattern across four of the five interviewees in describing their mentoring in terms of getting the pre-service teachers reflecting. The way the mentors articulated this aspect of their role was different in each case, however, so we take the four mentors in turn and briefly analyse each one.

> This year . . . I feel like I can advise because I just feel observing lessons I can see so much more in a lesson now than I would have done last year . . . there are so many questions I thought about in my research . . . Instead of just saying really black and white things which I felt I might have done more of last year, you know there was a kid doing this there was a kid doing that . . . I can be more like okay, how did you think that went? and ask slightly more challenging questions; instead of this was good this was good this was not so good, the questions which help them decide what was good and which was not so good and okay what could we have done differently. (Amy)

Amy has been in the role of mentoring a pre-service teacher for two years. She is aware of a significant shift taking place in her practice, from saying 'black and white things' about a lesson, and what was good or not so good, to supporting the pre-service teacher to figure out what went well, what did not go well and what could have been done differently. A similar practice is articulated by Sasha who has mentored for many years although was not in that role at the time of the interview (hence, we imagine, the past tense).

> I was always quite conscious of not trying to force someone into doing something they weren't comfortable with . . . allowing that person space to reflect on what they were doing and not always necessarily making suggestions.

> When you start working with a PGCE student on their first placement . . . there's too many things they have to learn, too many; so it's just deciding what's the one focus. (Sasha)

We see here Sasha describing what she does in a similar manner to the practice Amy is aware of moving into, i.e., not being the one making suggestions about what to do next or what went well, but rather supporting the pre-service teachers to come to these realisations. Sasha is aware of the complexity of what the pre-service teachers are experiencing; one focus is more effective than a list of things to change. In these comments, Sasha is explicit that she does not see her role as forcing the pre-service teachers to adopt any particular way of working.

A third mentor, Helen, commented:

> If I'm trying to encourage them to do something different and something exciting then I find that I've got to practice what I preach and OK we're going to do something different that's not just questions from the board or from a text book, doing something more interesting like YouTube or matching up puzzles, functional skills activities that I'm doing with them. It does encourage me to think more about how else I'm going to deliver a lesson to make it more creative, not necessarily more creative, it's to help them, the PGCE students understand the different ways of approaching a topic. It's sort of what I would know because I've done it a few different ways already. It's so that they can understand there're different ways that you deliver a lesson, they don't have to all be the same. It depends on the situation and what you want to do and depending on the students. (Helen)

What we are struck by in Helen's articulation of how she thinks about her work with pre-service teachers is how she, like Sasha, is not talking about trying to get them to adopt specific ways of working. Her attention is now on offering the pre-service teachers a range of practice and provoking in them an awareness that 'there're different ways that you deliver a lesson, they don't have to all be the same'. She is conscious now of deliberately exhibiting different approaches in her own lessons, when the pre-service teachers are observing. Helen commented on two pre-service teachers she was working with at the time exploring different methods when planning:

> I left them today . . . and they were still happily planning away together for their lesson tomorrow. They're doing the area of a circle and A was finding YouTube clips about area of circle with songs and B was drawing some big flow diagram on the whiteboard we've got in the room. Just 10 minutes earlier I'd seen her cutting up strips of paper whilst A was typing something else up then they were thinking about how to approach the topic of area of a circle whether they were going to talk about area of a circle or area of a square and almost 4 or 3 of them.

The fourth mentor, Adam, like Sasha has worked with PGCE students from the University of Bristol and other institutions for a number of years:

> I see my role as helping them supporting them in reflecting on what's happened, rather than telling. Er, probably more recently I have allowed more practical advice to be given, whereas I used to not do that at all really [laughter]. But it was then I had more time. And so, sometimes I guess I'm trying to hurry the process a little by 'try this next lesson and we can reflect on what happened'. I'm hoping the reflection is still important. But by suggesting try this then it might exaggerate the difference and make the reflection easier . . . I'm offering it in the hope there will be difference. (Adam)

Adam has recently taken on more responsibility in the department, hence presumably the comment 'then I had more time'. The way he describes his work uses the same language of 'difference' that we did above to describe our views about learning and researching. As with the other three mentors he describes 'supporting them in reflecting'. We see an intriguing cycle across the comments of these four mentors. Amy, at the start of her mentoring, is aware of moving to a position where she is more in the role of probing to get pre-service teachers reflecting, rather than telling them what to do. Adam is aware of having taken a more neutral and non-directive stance in the past and is now being more interventionist in terms of suggesting things to try, but his focus of attention is still on the reflecting of the pre-service teacher and his suggestions are designed to provoke difference (e.g., in this lesson compared to the last one) to aid this reflecting.

From learning to researching

So what does it mean to turn the mathematics learning environment (for students, pre-service or in-service teachers) into a research environment? Our response to this question arises theoretically, as a result of our beliefs about learning and researching, and practically, as a result of our analysis of interviews with school-based mentors. Both our theoretical convictions and experience suggest that it is possible to set up learning environments *from the start*, so that they are research environments. The phrase 'turning' the learning environment into a research environment perhaps suggests that the two kinds of environment are distinct and that one of them needs to change. Varela (1999: 32) writes about how beginners can be supported to learn (in any field) in the same way as experts if they steer a path between spontaneous action (which allows for no reflection) and over-deliberateness (that, in the complex space of the classroom, is too self-conscious to be effective). The middle way Varela (1999) advocates is a mode of being he labels 'deliberate analysis' (p. 27): the beginner is encouraged to act and then, afterwards, reconstruct the awarenesses that led to action (see Brown & Coles 2011 for more detail about these ideas and their application to learning to teach).

In the small group work of the PGCE course, pre-service teachers are forced to dwell on the detail of their experience (which is often hard for them at the start) in the way experts do, not jumping to interpretations or justifications until a more neutral description of events has been found. Similarly, in Alf's work on video, teachers are supported to act like experts by being forced to focus on the detail of events; this is what is involved in research and, yet, anyone can engage in the task. Expert and beginning teachers can act in the same way reconstructing the events on a small clip of a video recording of a lesson. The move to interpretation allows a reconstruction of our awareness of events on the video, by focusing on the strategies used by the teacher.

We see evidence in the mentor interviews that when teachers are used to researching their own practice (and, indeed, are supported in doing this from the very beginning of their journey into teaching), then it is natural to view their work with pre-service teachers as one of supporting a research journey. This may not be the language that is used explicitly, but the mentors describe giving space for the pre-service teachers to develop their own awareness of events and to decide on alternative actions for the future, a process we see as researching. We are fortunate on the University of Bristol PGCE course that a culture of learning to teach through researching teaching has developed over more than twenty years, so there are many mentors in schools who have also done the course.

In any mathematics learning environment there will be actions taking place and the potential for awareness *about* those actions (see Figure 11.2).

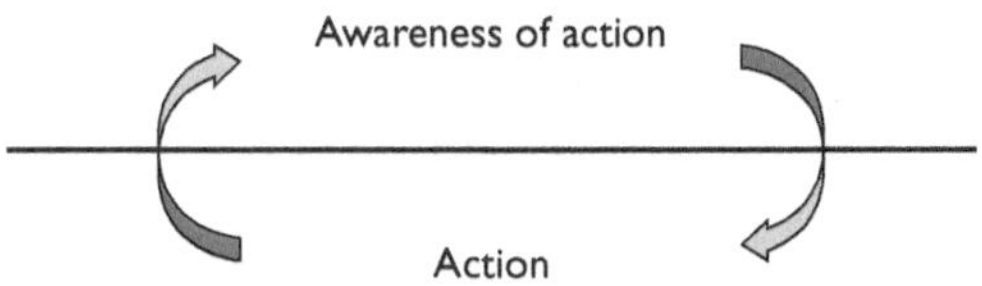

Figure 11.2 Action and awareness of action

Learning can take place in the realm of action. In the story at the start of this chapter, Laurinda prompts Sarah into acting differently (and therefore learning) without, at the stage we leave the story, an awareness *about* her actions (other than, perhaps, that she has done something different to earlier).

It is possible to give pre-service teachers templates for lesson planning, model lesson plans to follow, specific strategies to follow. All these elements are at the level of 'Action'. Learning can certainly take place. When Amy described her past work with pre-service teachers, when she told them what was good or bad, we would again interpret this talk as taking place at the level of action, and no doubt the teachers Amy was mentoring changed, developed and learnt. An approach to working with pre-service teachers that emphasises particular models of good practice could therefore be an example of a mathematical learning environment. But what such an approach to working with teachers may not provoke is an awareness *about* those actions or models of good practice. For example, the teacher may not have any strategies for what to do if the model or suggested action does not 'work' with a particular class or individual.

For us, what makes an environment a research environment is that an awareness of action is provoked, that cycles back to alter future action. Researching is a form of learning where there is awareness *about* that learning. The attention of the mentors we interviewed was firmly on the learning of the pre-service teachers. The mentors spoke of deliberately withholding or subordinating their own awarenesses of lessons, to the noticing of the pre-service teacher. In the PGCE course or work with video, there is no investment on the part of the tutor or facilitator about what teachers will notice and what purposes will arise; our attention is on creating the conditions in which it is likely that *some* purpose will arise. The pre-service teachers are therefore not placed in the position of trying to adapt to someone else's way of working without the complex awarenesses that make it successful. A beginner cannot act in the classroom in the same way as an expert does. What the beginner can do is be placed in an environment where they engage in reflecting in the same way as an expert does, i.e., researching. When this is successful there is a marked energy and commitment that accompanies the learning and the learning is fast.

Final remarks

The evidence we have presented in this chapter is that it is possible to work with pre-service teachers, from the beginning, to provoke awareness of the actions they are taking in the classroom. It is possible to set up, from the beginning, a research environment in learning to teach mathematics. The evidence from the interviews with mentors is that, if teachers have been involved in researching their own practice (from the beginning) then they talk of their work with other teachers not at the level of actions. In other words, the mentors are explicit about not offering models of practice, but supporting the pre-service teachers with whom they work to become aware of choices and possibilities and then reflecting on the consequence of those choices. These mentors, who have learned to teach by researching their practice, support pre-service teachers to learn to teach by researching their practice and extending the range of possible strategies.

References

Bateson, G. (1979) *Mind and Nature: A necessary unity*, Cresskill, NJ: Hampton Press Inc.

Brown, L. (2005) 'Purposes, metacommenting and basic-level categories: parallels between teaching mathematics and learning to teach mathematics', paper presented to the *15th ICMI Study Conference*. Online. Available HTTP: <http://stwww.weizmann.ac.il/G-math/ICMI/log_in.html> (accessed 28 January 2013).

Brown, L. (with Coles, A.) (1997) 'Being true to ourselves', in V. Zack, J. Mousley & C. Breen (eds), *Teacher as Researcher: Researcher as Teacher*, Victoria, Australia: Deakin University, pp. 103–11.

Brown, L. & Coles, A. (1997) 'The story of Sarah: seeing the general in the particular?', in E. Pehkonen (ed.), *Proceedings of the Twenty-first Annual Conference of the International Group for the Psychology of Mathematics Education*, Vol. 2, 113–20, Lahti, Finland.

Brown, L. & Coles, A. (2000) 'Complex decision making in the classroom: the teacher as an intuitive practitioner', in T. Atkinson & G. Claxton (eds), *The Intuitive Practitioner: On the value of not always knowing what one is doing*, Buckingham, UK: Open University Press, pp. 165–81.

Brown, L. & Coles, A. (2008) *Hearing Silence: Steps to teaching mathematics*, Cambridge: Black Apollo Press.

Brown, L. & Coles, A. (2011) 'Developing expertise: how enactivism re-frames mathematics teacher development', *ZDM Mathematics Education*, 43(6–7), 861–73.

Brown, L. & Dobson, A. (1996) 'Using dissonance – finding the grit in the oyster', in G. Claxton, T. Atkinson, M. Osborn & M. Wallace (eds), *Liberating the Learner: Lessons for professional development in education*, London: Routledge, pp. 212–27.

Brown, L., Sutherland, R., Winter, J., & Coles, A. (2001) *Final Report: Developing algebraic activity in a 'community of inquirers'* (Document Number R000223044), Swindon: ESRC.

Claxton, G. (1996) 'Implicit theories of learning', in G. Claxton, T. Atkinson, M. Osborn & M. Wallace (eds), *Liberating the Learner*, London: Routledge, pp. 45–56.

Coles, A. (2012) 'Using video for professional development: the role of the discussion facilitator', *Journal of Mathematics Teacher Education*, online first, DOI 10.1007/s10857-012-9225-0.

Coles, A. (2013) *Being Alongside: For the teaching and learning of mathematics*, Rotterdam: Sense Publishers.

Jaworski, B. (1990) 'Video as a tool for teachers' professional development', *Professional Development in Education*, 16(1), 60–5.

Mason, J. (2002) *Researching Your Own Practice: The discipline of noticing*, London: RoutledgeFalmer.

Nemirovsky, R., Dimattia, C., Ribeiro, B., & Lara-Meloy, T. (2005), 'Talking about teaching episodes', *Journal of Mathematics Teacher Education*, 8(5), 363–92.

van Es, E. & Sherin, M. (2008) 'Mathematics teachers' "learning to notice" in the context of a video club', *Teaching and Teacher Education*, 24, 244–76.

Varela, F. (1999) *Ethical Know-how: Action, wisdom and cognition*, Stanford, California: Stanford University Press.

Chapter 12

Reflective practitioners for lifelong learning

Julie-Ann Edwards and Ray Huntley

Reflective practice in teaching and learning is considered to be one of the most effective means through which personal and professional development occurs. An increasing literature base links the quality of professionals in training with their capacity to reflect critically, accurately and constructively on their practice. This chapter explores how such reflection allows teachers to constantly react to and review their experiences in order to reformulate their practice. We use ideas contained in the *Discipline of Noticing* to exemplify how reflective practice can be utilised to improve practice and draw on our research evidence from pre-service and experienced teachers who reflect on how professional development experiences have impacted on their practice of teaching and on their students' learning.

Introduction

A wealth of literature confirms that to be a reflective teacher is a positive attribute. Brown and McNamara (2011) report on a range of research evidence which tracks the growth in expectation of both beginning and experienced teachers to develop reflective practice as an integral part of professional development. Drawing on work on teacher identity, they suggest that "reflective work provides a forum in which students [pre-service teachers] can reconstruct their own identity as they become inducted into professional discourses" (2011: 39–40). This is akin to Donald Schön's argument, mentioned by Mark Simmonds and Jenni Back in Chapter 10, that teachers develop and integrate reflective skills as they gain experience. Schön's model of moving from knowing-in-action, through reflection-in-action, to reflection *on* reflection-in-action provides a theoretical lens through which the reflective development of teachers can be viewed.

A theoretical lens on reflection

It might be argued that a competency-based assessment of pre-service teachers could lead to what Schön terms 'technical rationality' (1991: 21). He describes this as an instrumentalist approach to professional practice, one we might see early on in the practice of pre-service teachers who abide by a rule-bound means through which they meet the end outcome of teaching. Schön argues that technical rationality leads to a loss of *craft knowledge* and *artistry* in professional practice. Ward and McCotter (2004) also suggest that a move to assessing pre-service teachers not only on their skills for teaching, but also on their impact on students' learning (Ofsted 2008; Mitchell 2000) in both the UK and US settings, requires a different sort of reflection on and in practice. The need, therefore, is for initial teacher education training processes to develop a model of reflective practice as a means of encouraging pre-service teachers to step far beyond technical rationality as soon as possible.

Schön's vision of *professional artistry* (Schön 1983, 1987) is described as the "kinds of competence practitioners sometimes display in unique, uncertain, and conflicted situations of practice" (Schön 1987: 22). He argues that professionals can act at three levels of reflective activity. The first of these is 'knowing-in-action' which is demonstrated in the skilled physical actions undertaken by teachers in practice in the classroom. Schön's focus on dynamic activity, as a crucial element of professional *artistry*, is exemplified in his distinction between *knowing-in-action* and *knowledge-in-action*. The former represents the active state of doing the individual skilled actions that make up skilful teaching without necessarily knowing that these are being done or being able to verbalise the means by which these are done; "[The] knowing is *in* the action" (Schön 1987: 25). If this *knowing-in-action* is able to be described, it becomes the passive form of *knowledge-in-action*.

Schön calls the second level of reflective activity *reflection-in-action*. He describes this as the capacity of professionals to consciously think about what they are doing while they are doing it. As a pre-requisite for reflection-in-action, he envisaged the competence of teachers and the *artistry* (or creativity) in using this competence to be already embedded in skilled practice. This is necessary to enable the teacher to 'think on their feet' in response to an unusual event and take appropriate action either from the repertoire of skilled knowledge-in-action or by inventing a new solution. As with knowing-in-action, teachers may not be aware of their reflection-in-action, even if it results in further knowing-in-action.

The third level of Schön's model utilises ongoing iterations of reflection:

> Clearly, it is one thing to be able to reflect-in-action and quite another to be able to reflect on our reflection-in-action so as to produce a good verbal description of it; and it is still another thing to be able to reflect on the resulting description. But our reflection on our past reflection-in-action may indirectly shape our future action.
>
> (Schön 1987: 31)

It is this reflection on reflection-in action (or meta-reflection) which is the aim in increasing teachers' capacity to proactively improve teaching and learning.

Reflecting as a pre-service teacher

Over the years, much has been written about the means by which pre-service teachers can be encouraged to reflect effectively on their teaching (see, for example, Ghaye & Ghaye 1998; Pollard 2005; Wildman, Niles, Magliaro, & McLaughlin 1990). Evidence for such reflection comes in many forms, some of which are identified here:

- post-formal lesson observation discussion with the mentor or class teacher about a lesson;
- informal discussions;
- post-teaching annotation of lesson plans (whether formally observed or not);
- written accounts of lesson evaluations;
- targets for areas for improvement, with justifications for these decisions;
- formally assessed written elements of an initial teacher education programme;
- written accounts undertaken as part of a research project.

Most of this evidence will be highly visible to the practice-based mentor. As part of the usual process of mentoring, a debrief following an observed lesson, whether this is informal or formal, will inform mentors of the extent to which a pre-service teacher is able to accurately evaluate the lesson at the levels of their teaching skills, the content and the students' learning. Reflection may form part of this evaluative process if the pre-service teacher is able to consider how the lesson evaluation impacts on future practice. Hence a practice-based mentor's question, "What would you do differently?" is an important trigger for encouraging reflective practice during a lesson debrief or a discussion about written accounts of lesson evaluations. Similarly, encouraging pre-service teachers to actively annotate lesson plans, following reflection on the taught lesson, is a practice which will aid them to maintain an immediately accessible means to improvement of a similar future lesson. Regular meetings between the practice-based mentor and the pre-service teacher in a mentoring relationship will offer a wealth of opportunity for developing effective reflection for both the pre-service teacher and the mentor.

What may seem invisible to the practice-based mentor are the final two forms of reflection listed above. These generally take place under the auspices of the accrediting provider institution. However, as Brown and McNamara (2011) point out, there is an increasing trend in expectations of pre-service teachers that they engage in activities such as action research, micro-teaching and writing reflective logs to increase the range of opportunities in which pre-service teachers can engage overtly in reflective activity. We add design research (Kelly, Lesh & Baek 2008), to this list, as an activity which enables pre-service teachers to focus on reflection for improving engagement with central curricular issues related both to teaching and students' learning. These activities then become much more part of the practice-based mentoring collaboration.

As described by Mark Simmonds and Jenni Back in Chapter 10, and by Alf Coles and Laurinda Brown in Chapter 11, lessons which are videoed can provide a wealth of opportunities for collaborative professional dialogue. They also establish a basis for effective reflection, as the video can be viewed repeatedly. Our own experience is that micro-teaching a 20-minute session with a specific curricular focus, very early in the initial teacher education programme, enhances the capacity of pre-service teachers to reflect more accurately on their teaching skills throughout the programme. More recently, and with the increasing accessibility of appropriate technology, these micro-teaching sessions are being videoed, using Flip digital cameras. Each recording is sent to the pre-service teacher for use either in personal reflection or as a focus for mutual peer reflection. Similar scenarios can be generated in practice-based settings, with a focus on the pre-service teacher teaching, or on a group of children learning, as an evidence base for discussion between the pre-service teacher and mentor about teaching and learning.

Evidence of reflection

In a study of a formally assessed element of primary pre-service teachers' writing (Edwards 2007) across two cohorts on a one-year postgraduate initial teacher education programme, we found that these pre-service teachers' reflections on the impact of undertaking small-scale research in their practice-based settings fell into four categories:

- reflection on outcomes of the classroom-based research undertaken and how this directly impacts on the pre-service teacher's views of teaching and learning;
- reflection on professional and personal development more generally;
- evaluation of the impact of the research outcomes on the practice-based mentor or wider school context;
- confirmation of beliefs or understandings about theoretical perspectives.

The following excerpts from evidence collected from pre-service teachers' writing indicate each of the categories above.

Reflection on outcomes:

> I was surprised by the lack of directive scaffolding the children required in the discussion group to remain on task and develop their thinking. I had previously regarded [my] scaffolding as an extensive and dominant element of the discussion. I now realise that in a collaborative setting in which talk 'rules' and understandings are firmly established, scaffolding is more subtle than this, guiding and being responsive in a child-focused way.

Reflection on professional and personal development:

> I appreciate better now than before the value of carrying out such tasks, so as to improve my own teaching methods and eventually the learning of the

> children I teach . . . I have now become more mindful of the benefits of research on a regular basis, to 'refocus' one's teaching, and to gain a better understanding of the approaches that work best in one's own practice and those which should be reconsidered or 'reworked'.

Evaluation of impact of outcomes on the wider school context:

> On reflection, the class teacher thought she had gained awareness that the children needed to be taught how to talk together. Due to the positive attitude and learning the children had displayed in the sessions, she has decided to use the talk rules. She had not previously realised the percentage of teacher talk that naturally occurred in the classroom and recognised the need for the children to be more proactive and be given the opportunity to assume the responsibility they had so successfully adopted.

Beliefs about theoretical perspectives:

> The research has reaffirmed my belief in the social-constructivists' claim that language is used in the process of thought and is not merely a product of it.

> Although it is beneficial to read about theories, putting these theories into practice gave me a real understanding of these issues.

As a mentor of pre-service teachers, looking for elements of these types of reflection in discussions, written accounts and target-setting for development, and making them overt to a pre-service teacher, can help to develop the skill of meta-reflection – what Schön termed reflection on reflection-in-action – and thus encourage a depth to their analysis of practice.

A model for pre-service teacher's reflection

In a study in which Ward and McCotter (2004) similarly analysed written accounts of pre-service teachers' reflections on the assessments of students during a series of lessons, they identified four qualitatively different levels of reflection: routine, technical, dialogic and transformative. They also used three aspects of reflection to further quantify the quality of reflection: focus, inquiry and change. The focus aspect showed how pre-service teachers accounted for concerns about their practice. The inquiry aspect indicated how pre-service teachers questioned their practice, while the change aspect signified how each of the previous two aspects impacted on how pre-service teachers changed their practice and perceptions of teaching and learning. These elements are summarised in Table 12.1.

Experienced mentors will recognise these categories in many of those whom they have mentored. For example, a pre-service teacher acting within the Routine/Focus category may evaluate a lesson which has not gone according to

Table 12.1 Framework of categories of reflection in pre-service teachers (after Ward and McCotter 2004)

	Routine	*Technical*	*Dialogic*	*Transformative*
Focus	Centred on self, behaviour management and achieving accreditation.	Centred on teaching tasks and assessment which is functional for the pre-service teacher's needs.	Centred on student's learning, particularly low-attaining students.	Centred on pedagogical issues which are affected by wider socio-cultural dimensions.
Inquiry	Unquestioning about self with a focus on accounting externally for teaching issues.	Questions relate to specific situations and cease when the problem is solved.	Questions openly consider the views of all participants, are collaborative and developmental.	Questions of self are challenging and high-level with a critical approach to experiences.
Change	Role of evaluation of practice is functional.	Response is personal but does not result in change.	New insights about self, teaching and students' needs.	Reformulating practice on the basis of reflections.

plan by saying, "There was a group which was not behaving, so the lesson didn't go well." A similar lesson might be evaluated by a pre-service teacher in the Technical/Inquiry category by a question such as "How can I make sure that the disruptive group is engaged in the task?" A pre-service teacher in the Dialogic/Change category might state that "I didn't differentiate enough for the group which didn't engage with the task." The challenge for the practice-based mentor is to assess where each of the pre-service teachers with whom we work is located in this framework, in order to act accordingly and most effectively for their improving reflective practice. However, such a model of reflection is not necessarily confined to pre-service teachers. More experienced teachers might equally be located in any of these categories. We now propose a model through which experienced practice-based mentors might engage with reflection on practice in a way which moves them closer to the Transformative/Change category.

Reflecting as research

Within the field of practitioner research and reflective practice, one method has been identified as being of significant value in the challenge to make sense of the messiness experienced by teachers researching their own practice. Mason (2002) describes in his book *Researching Your Own Practice: The Discipline of Noticing* a method for gathering data from one's teaching in the form of momentary acts of noticing and concise accounts of practice and then using these as a basis for reflection with the aim of improving practice through considering how a teacher might act when faced with similar situations in their future teaching. The National Council for Teachers of Mathematics (2010) attaches great value to studies which adopt such approaches for professional development, identifying four core goals for mathematics teachers engaged in professional development programmes:

- to build teachers' knowledge of mathematics and their ability to use it in practice;
- to build teachers' capacity to notice, analyse and respond to students' thinking;
- to build teachers' productive habits of mind;
- to build collegial relationships and structures that support continued learning.

These goals themselves build on what Ramsden (1992: 5) argues when describing "a reflective and enquiring approach as a necessary condition for improving teaching", as well as believing that constant practice and reflection improves understanding of how to use the range of skills employed in teaching. One of the critical roles of a practice-based mentor is to model and reinforce this approach with pre-service teachers.

The discipline of noticing

Eisner (1985) argued that in order for educational practice to be fully and truly informed by educational research, there is a need for teaching practitioners to "construct their own unique conceptual apparatus and research methods" (1985: 264). In pursuit of this goal, Mason developed the Discipline of Noticing which sets out to provide reflective practitioners with foundations on which they can research 'from the inside', moving from having a series of teaching experiences to researching how to change personal practice in order to improve learning experiences for students. In this way, teachers should not be simply residing in their practice as a professional, working to improve that practice, but actively researching that practice to make it better in ways which may not appear obvious until research uncovers them.

The first aspect of the Discipline of Noticing to which we now draw attention is that of physiology. Mentors might like to consider some of the following elements of their position, posture or actions and how these might influence behaviour whilst reading this book or whilst going about everyday practice as a teacher. For instance, notice how feet are positioned, notice which nostril(s) are being breathed through. Mentors might consider that noticing these aspects is not really going to matter in the grand scheme of professional development but, as attention starts to be drawn to things such as these physiological items, things which are normally only vaguely part of awareness at the best of times, a sensitivity will begin to develop to notice specific things which, in turn, inform us of something about ourselves. When we later begin to notice the finer details of our practice, it will enable us to sharpen our practice in very particular ways, over which we have more control once the noticing is developed sufficiently.

In the classroom, mentors could consider how different postures, gestures, expressions and tones of voice reflect their beliefs as a practitioner, perhaps when asking a question to find out something not currently known or when looking for particular answers. At other times, when engaged in 'being a teacher', try to notice where attention is drawn. For example, what is being attended to when listening to a struggling student trying to explain their concerns? What about when working through an example with a group of students? To what is attention drawn when preparing a lesson? Once a mentor is aware of these elements of practice, they can be usefully shared with pre-service teachers. It may be that, in the course of practice as a teacher, a mentor habitually adopts a particular expression or voice tone in certain situations that can signal to a student that they are pleased, distant, caring or confrontational to some extent. By learning to notice such aspects, in the moment just before adopting them, the teacher is then able to effect a change from habit in order to demonstrate to the pre-service teacher the kind of approach being made at that time.

The next key principle for the Discipline of Noticing is that of being able to 'say what you see'. Mentors might like to try a brief experiment for themselves and with beginning teachers – find a drawing of a shape or simple pattern, stare at it for a short while, then try to describe it to yourself. Then write some instructions on how to draw it, which can be given to someone else. When you describe what you see to yourself, it is more likely that you will recreate the image quite accurately in your own mind. However, when describing it to someone else, either verbally or in written instructions, the other person may visualise something rather different to what you see. This is also true of describing episodes from a teaching session. Your description of what you think took place may be meaningful for you, but when you describe it to someone who wasn't present in the lesson, they may imagine a very different scene. In order to try to ensure that any recall of some teaching was accurate, the process of making accounts should be utilised.

Accounts-of and accounting-for

The purpose of making an account is to achieve clarity of description of an event so that anyone analysing and discussing it can have the best possible picture of what took place. Such a description should necessarily only focus on pure description and not contain anything that might appear to be evaluative, otherwise the analysis of the event from the description could be biased. To this end, Mason (2002) sets out two types of account of an event, each containing distinctive elements of the observed session.

The first is known as the 'account-of', which is a brief but vivid description of exactly what took place, but without evaluation, analysis, explanation or justification of what happened. Such an account contains only those details that could be agreed by other observers who were present at the time. This type of account allows the writer access back to the situation and allow readers similar access to similar situations in their own experience. A brief example of this might be as follows: "The teacher led a starter activity in which he displayed a picture of a cheetah on the screen and asked the children to estimate how many spots could be seen on the cheetah. One child immediately responded by saying there were hundreds, and the teacher asked how the child knew that. The child laughed and altered the answer to millions, at which point the teacher asked the group again and chose another child to respond." In this example, the teacher clearly seems to be unsure how to deal with the first child's response, so opts instead to ask another child. Any observer in the room could verify the narrative of what occurred, without reference to judgements about what the teacher might have done, or analysing why he acted as he did. However, using this account, the incident could be recreated for discussion and analysis. This becomes the 'account-for', in which the description is a basis for theorising, explaining and analysing what happened, as well as perhaps indicating some of the emotional aspects of the experience. Taking the example above, it might be surmised that the teacher could have targeted the question to particular children. Strategies for dealing with the 'silly' response could perhaps be identified in a more detailed account. It is the former account-of that is regarded as being most helpful when seeking a focus for mutual reflective practice and, thus, for collective professional development.

Intentional noticing

One of the ways in which a novice teacher can begin to develop a sense of their own practice is to engage in what is described as noticing. This could include all manner of sensual experiences – sights, sounds, smells and various emotions. Noticing these is termed 'ordinary noticing' and is a useful first step. With greater realisation of what they have noticed, teachers might begin to internalise this through casual commenting to a colleague, friend or mentor; for example, a teacher may wish to remark that a particular element of a lesson went well, with some reason given, perhaps that the activities kept the children engaged for longer than expected. Mason (2002) describes this as 'marking', being a more collaborative and interpersonal experience, and that this may trigger a response, or remark which can give us fresh access to our own experience. However, to have a significant effect on reflective practice, a teacher should develop 'intentional noticing', something which does not happen automatically for most people. It requires teachers to sensitise themselves to noticing and to see their own practice through new eyes. Examples of this can come from a range of contexts. Perhaps, as a mentor, suggestions can be made to a novice teacher about focusing on the clarity of their questions and responses by repeating back a response given by a child. Tuning in to see if this is occurring and making a note, mental or otherwise, of whenever it happens, and perhaps when it doesn't, is an important aspect of noticing. In discussing the lesson with the novice teacher at a later point, inviting them to comment on and recall how effective they were in repeating responses, can help the teacher develop their own 'intentional noticing'.

As an experienced teacher, a practice-based mentor can consider an incident, situation or phenomenon in their own classroom. The first aspect to noticing what happened is to recognise an issue and an opportunity to act. Consideration of what might be done differently in future, if a similar situation occurs, leads to the opportunity to 'act freshly', providing support for further reflection and action, through which sensitisation to notice occurs more strongly in the future.

With novice teachers, this cyclical action can be adopted to help them develop the skills of intentional noticing. They will often be keen to make the new and unfamiliar become part of their regular and familiar routine – moving from Schön's (1987) knowing-in-action to reflection-in-action. Using competence here to mean being able to do something well, a state of flow is generated between states of competence. Once we have intentionally noticed that we have slipped into unconscious incompetence, a cycle of reflection and improvement follows, through conscious incompetence to conscious competence, and back to unconscious competence.

Developing from experience

The 'account-for' scenario, where the experience of attempting to explain and justify effectively the description outlined in an account, derives a narrative of the experience that is useful in helping us to justify that experience. It is well established that we learn from experience; for example, Kolb (1984) developed a model of experiential learning, based on a number of propositions:

- learning is a process, not a series of outcomes;
- learning is about refining previous ideas and beliefs;
- learning requires resolution of differences of perception;
- learning is the integrated functioning of the whole person;
- learning occurs through synergetic transactions between learner and environment;
- learning is the process of creating knowledge.

When mentoring a novice teacher, it is essential to enable that teacher to engage with these propositions through experiencing, reflecting, thinking and acting, in an iterative process of development, in the ways we have described earlier.

Experienced teachers reflecting on teaching

One of the most important benefits of being a reflective practitioner is that it provides opportunities to modify and extend practice as either a beginning or an experienced teacher so that the consequence is students' enhanced learning experiences. By reading, hearing about and observing others teaching, practice-based mentors and beginning teachers can collect ideas and try out, within their own teaching situations, versions of what was read, heard about or observed. Gathering information from episodes in others' classrooms enables pre-service teachers, in particular, to begin to change themselves, always seeking to become more sensitive to the needs and learning experiences of their students. A teacher of mathematics should, and probably would, be continually engaging with mathematics, working on mathematical problems, keeping skills sharp and always looking to learn something new through a topic or a problem to keep the subject matter interesting, challenging and refreshing for both themselves and their students. Such an approach can be modelled to beginning teachers to enable them to develop good habits in relation to mathematics learning. Jaworski and Matthews (2011) used a seminar series in a university-based School of Mathematics to focus on examining how mathematicians and mathematics educators, based in the School, talked about their teaching, rather than examining how mathematics is taught in the School. In seeking such a discourse on teaching, the focus was not on practice, as such, but on how the mathematics community shared its thinking about the design and practice of teaching within the institution. This is a model which is easily transferable to other practice-based settings for mentors and beginning teachers. Similarly, research evidence from the Community for Undergraduate Learning in the Mathematical Sciences (CULMS 2009) shows how collaborative reflection on teaching also drives improvement in teaching and learning.

Experienced teachers reflecting on professional development

Seven mathematics teachers, who are practice-based mentors in secondary schools, participated in a research project during their study for an MSc in mathematics education part-time over two years (Edwards 2008). They recorded what they perceived to be significant shifts in their thinking about the teaching and learning of mathematics. Evidence suggested that transformative reflection and subsequent change could be categorised into five forms of impact:

- personal impact;
- professional impact;
- impact on the practice-based mentor's classroom;
- impact on the practice-based mentor's department or school;
- impact beyond the school.

The first of these categories indicates development of self-knowledge (personal) rather than development of knowledge which has an overt impact on aspects of professional life and is indicated by the following two quotes from mentors:

> Group work is messy and I'm quite tidy, so I was definitely out of my comfort zone to have children not facing me. I was surprised about how insecure it made me feel and I realised how much I needed to be in control in my classroom. After all these years teaching, I'm learning new things about myself.

> I think "it doesn't matter what their books look like as long as they're thinking" until I realise I'm thinking "what do their books look like?" I think I say I believe all these things but I've found I don't actually do this in practice. I was reading something about the [inconsistent] relationship between beliefs and practice and I thought "Oooh, that's me".

Interactions with beginning teachers may provoke similar reactions in an experienced mentor – and these can be acknowledged, shared and 'unpicked' in discussion as a mutual learning platform.

The examples of impact on professional practice were very numerous but we provide two examples here:

> Reading Geoff Wake's [2005] article prompted me to think about the way that I approach topics. I plan to think more carefully about *beginning* a topic with a problem and asking "what do we need to know or learn to solve this?"

> Problem-solving ideas are much more in my thoughts when lesson planning now – I'm aware of a shift in my previous thinking.

Similarly, the impact on mentors' classrooms provided prolific examples, of which two are offered here:

> I have begun starting all topics with problems. A good introduction to [class 10]; linking Pythagorean triples with the radius of a circle means they are now all well-acquainted with Pythagoras.

> I never believed in using friendship groups in the classroom because I thought they'd [the students] just chat but, when I listened to the audio tapes, I was really surprised at the level of on-task talk for an hour – 93%.

The ideas which beginning teachers bring to a specific teaching setting may stimulate new thinking amongst colleagues, in a way similar to the external stimulus of a Masters degree course. Pre-service teachers, in particular, tend to have additional time and energy to resource lessons creatively. If they are given the space and freedom to do so, within a construct of ensuring effective learning for the students they are teaching, they can provide a stimulus for curricular change for a mentor, a department or a school. The following two excerpts from the research study are indicative of how a pre-service teacher's work for the provider institution might impact on change within the practice-based setting through sharing experiences in a collaborative mentoring relationship:

> Will get a copy of Prestage and Perks [2001]. Very interesting reading – have made a list of suggested approaches and good ideas to add to our [department] schemes of work.
>
> Borrowed the DIME Build-ups [Tarquin] and used some department time on our INSET [in-service training] day to introduce them to the staff. We really enjoyed them!!

Our emphasis on collaboration in the mentoring role is an important element of our understanding of how reflective practice promotes longer-term learning for experienced teachers in the context of teaching and learning.

Reflective practice as a lifelong learning journey

In developing a scheme for promoting reflective practice in experienced teachers that enabled teachers to feel more intellectually involved in their role and their work in teaching and to cope with the lack of specific definitions for *craft knowledge* in the discipline of teaching, Wildman and Niles (1987) identified three features which facilitate reflective practice in teaching situations:

- support from administrators, enabling teachers to understand the requirements of reflective practice and how it relates to teaching students;
- time and space being sufficient to put aside the immediate demands of students in order for teachers to develop reflective skills;
- development of a collaborative environment with the support of other teachers.

These features will be present in settings where lesson study or other mutual forms of collaboration, described in Chapter 6, are part of the community of practice of a school or mathematics department. The professional practice of teaching and the expectations of learners are likely to continue to be subject to the vagaries of external influences, particularly political. The constant ebb and flow of change in educational settings and expectations means that the teaching profession requires critical and adaptive professionals, who understand that teaching is a lifelong journey of professional and personal learning. Central to these critical and adaptive requirements are practice-based mentors, who not only model but also pro-actively practice, reflection on their reflection-in-action.

References

Brown, T., & McNamara, O. (2011) *Becoming a Mathematics Teacher*, Dordrecht: Springer.

Community for Undergraduate Learning in the Mathematical Sciences (2009) 'Mathematics undergraduates teaching: Perspectives and interactions project'. Online. Available at HTTP: <http://www.math.auckland.ac.nz/CULMS/projects/> (accessed 28 December 2012).

Edwards, J. (2007) 'Primary trainees' reflection-in action', *Proceedings of the British Society for Research into Learning Mathematics*, 27(1), 30–5.

Edwards, J. (2008) 'The impact of Masters level study on teachers' professional development', *Proceedings of the British Society for Research into Learning Mathematics*, 28(3), 30–5.

Eisner, E. (1985) *The Art of Educational Evaluation*, London: Falmer.

Ghaye, A. & Ghaye, K. (1998) *Teaching and Learning through Critical Reflective Practice*, London: David Fulton Publishers.

Jaworski, B. & Matthews, J. (2011) 'How we teach Mathematics: Discourses on/in university teaching', *Proceedings of the 7th Congress of European Research in Mathematics Education*. Online. Available at HTTP: <http://www.cerme7.univ.rzeszow.pl/WG/14/CERME7-WG14-Paper--Jaworski-&-Matthews-REVISED-Dec2010.pdf> (accessed 10 November 2012).

Kelly, A. E., Lesh, R., & Baek. J. (eds) (2008) *Handbook of Design Research Methods in Education: Innovations in science, technology, mathematics and engineering learning and teaching*, New York: Routledge.

Kolb, D. (1984) *Experiential Learning: Experience as the source of learning and development*, Englewood Cliffs, NJ: Prentice Hall.

Mason, J. (2002) *Researching Your Own Practice: The discipline of noticing*, Abingdon: Routledge.

Mitchell, A. (2000) *NCATE 2000: Teacher Education and Performance-Based Reform*, US: National Council for Accreditation of Teacher Education.

National Council for Teachers of Mathematics (2010) *Professional Development Research Brief*. Online. Available at HTTP: <http://www.nctm.org/uploadedFiles/Research_News_and_Advocacy/Research/Clips_and_Briefs/Research_brief_15-Goldsmith(1).pdf> (accessed 28 December 2012).

Office for Standards in Education (2008) *Mathematics: Understanding the score*, Report no. 070063, London: Ofsted.

Pollard, A. (2005) *Reflective Practice*, London: Continuum.

Prestage, S. & Perks, P. (2001) *Adapting and Extending Maths: New tasks for old*, London: David Fulton Publishers.

Ramsden, P. (1992) *Learning to Teach in Higher Education*, New York: Routledge.

Schön, D. (1983) *The Reflective Practitioner*, New York: Basic Books.

Schön, D. (1987) *Educating the Reflective Practitioner*, San Francisco: Jossey-Bass Inc.

Schön, D. (1991) *The Reflective Practitioner: How professionals think in action*, Avebury: Ashgate Publishing Ltd.

Tarquin Group (n.d.) *DIME Build-ups*. Available at HTTP: <http://www.tarquingroup.com> (accessed 18 February 2013).

Wake, G. (2005) *Functional Mathematics: More than "back to basics"*, Nuffield Review of 14–19 Education and Training: Aims, Learning and Curriculum Series, Discussion Paper 17.

Ward, J. & McCotter, S. (2004) 'Reflection as a visible outcome for pre-service teachers', *Teaching and Teaching Education*, 20, 243–57.

Wildman, R. and Niles, J. (1987) 'Reflective teachers: tensions between abstractions and realities', *Journal of Teacher Education*, 3, 25–31.

Wildman, T., Niles, J., Magliaro, S., & McLaughlin, R. (1990) 'Promoting reflective practice among beginning and experienced teachers', in R. Clift, W. Houston, & M. Pugach (eds) *Encouraging Reflective Practice in Education*, 139–62, New York: Teachers College Press.

Index

The word order is letter by letter; locators to plans and tables are in *italics*.